Narrative as Virtual

Narrative as Virtual Reality 2

Revisiting Immersion and Interactivity in Literature and Electronic Media

Marie-Laure Ryan

JOHNS HOPKINS UNIVERSITY PRESS   BALTIMORE

 Published 2015
Printed in the United States of America on acid-free paper
9 8 7 6 5 4 3 2 1

Johns Hopkins University Press
2715 North Charles Street
Baltimore, Maryland 21218-4363
www.press.jhu.edu

Library of Congress Cataloging-in-Publication Data

Ryan, Marie-Laure, 1946–
Narrative as virtual reality 2 : revisiting immersion and interactivity in literature and electronic media / Marie-Laure Ryan.
pages cm
Revised edition of: Narrative as virtual reality : immersion and interactivity in literature and electronic media. 2001.
Includes bibliographical references and index.
ISBN 978-1-4214-1797-4 (pbk. : alk. paper) — ISBN 978-1-4214-1798-1 (electronic) — ISBN 1-4214-1797-9 (pbk. : alk. paper) — ISBN 1-4214-1798-7 (electronic) 1. Books and reading. 2. Hypertext literature. 3. Electronic games. 4. Interactive multimedia. 5. Virtual reality. 6. Narrative (Rhetoric) I. Title. II. Title: Narrative as virtual reality two : revisiting immersion and interactivity.
Z1003.R97 2015
028'.9—dc23 2015002502

A catalog record for this book is available from the British Library.

Special discounts are available for bulk purchases of this book. For more information, please contact Special Sales at 410-516-6936 or specialsales@press.jhu.edu.

Johns Hopkins University Press uses environmentally friendly book materials, including recycled text paper that is composed of at least 30 percent post-consumer waste, whenever possible.

Contents

Figures and Tables

FIGURES

TABLES

Preface

One of the countless television reality shows running at the time of this writing (*Love It or List It*) features people whose lifestyle no longer fits their current home. Will they sell or will they renovate? Most of the time they renovate because it is more economical and because they are emotionally attached to the old place. Similarly, when the prospect of doing a second edition of *Narrative as Virtual Reality* came up, I faced the dilemma of writing a whole new book or of flipping the old one. Simply reprinting the original text was not an option because the cultural and technological environment with which this book is dealing has dramatically changed since the original publication. The most important of these changes is the disappearance of virtual reality (VR) from the radar of the media. This does not mean that the technology has stayed put since the turn of the century, but its developments have taken place in highly specialized scientific or military domains rather than in areas accessible to the general public, such as the entertainment industry. Hence the loss of media coverage, though Mark Zuckerberg's 2014 purchase of Oculus Rift, a company that makes lightweight HMDs (head-mounted displays) usable in video games, and the ongoing talk about the release of Google glasses may restore some of the public attention that VR formerly enjoyed.

For the literary scholar, the main difference between then and now is the loss of the prominence of hypertext. When I wrote the first edition of this book, the digital revolution had just reached the humanities, and hypertext fiction seemed on the verge of dominating the digital future of literature, if not the future of narrative, period. Much of the book was written in a dialogue with hypertext theory; now hypertext is no longer perceived as *the* narrative use of digital technology but only as one of many possible such uses. Therefore, the focus had to be altered in order to take into consideration the many other kinds of digital narratives that have emerged in the meantime. Another factor of change is the waning of the influence of what I call "textual" theories of literature, meaning by this theories, such as New Criticism or deconstruction, that conceive the literary text as a play of signifiers and regard its language as essentially self-referential. Now that it is acceptable again to talk about content, mimesis, reference, emotional involvement with characters, and immersion in fictional worlds, the occasionally polemical tone of the confrontation of the original text with textualism could be muted.

Because of these climactic changes, I had to renovate the book. I took down walls between chapters, fusing the former chapters 4 and 5, created new spaces (6), added content, moved around some passages, and relegated the "interludes" of the first edition (discussions of specific texts that accompanied the theoretical chapters) to an accompanying Web site. The revisions begin slowly. I made only minor changes to chapter 1, which deals with the philosophical notion of virtuality, a largely atemporal topic. Similarly, chapter 2, which deals with the technological notion of VR, has not been significantly modified because it is not concerned with actual implementations but with the dreams that motivated early developers. More substantial changes begin with chapter 3. The decline of the textual theories mentioned above has been counterbalanced by cognitive approaches to literature and narrative which provided new insights on the mental experience of immersion. Keeping up with this cognitive turn, the rewriting of chapters 3 and 4 integrates recent work on mental simulation, emotions, and empathy for fictional characters.

Chapter 5, which opposes the game aesthetics to the world aesthetics has undergone only minor changes, but a whole new chapter has been added to deal with an important trend in electronic literature: the rejection of narrative world creation in favor of play with language and the digital medium that questions their functionality. The chapter on forms of interactivity adds to its survey of interactive textual architectures a taxonomy of modes of user participation that runs from the external/exploratory mode of hypertext to the internal/ontological mode of video games. While the chapters on hypertext and on nondigital forms of user participation reorganize much of the material presented in the first edition (with some additions for the latter), the last chapter, on the problematic integration of user agency into immersive narrative schemata, shares only a title and a concern with the earlier version. It is based on my experience with a vastly expanded range of digital texts, especially computer games, and it takes a much more direct and constructive approach to the design issue. The conclusion is entirely new; it surveys the digital narrative genres that have been born, and the theoretical work of artificial intelligence (AI) specialists on interactive storytelling that has taken place since the first edition of this book.

Just as architectural remodeling must respect the foundations of a house, this revised version of *Narrative as Virtual Reality* maintains the basic thesis of the original: that disembodied, external interactivity is hostile to immersion, and that the fullest reconciliation of interactivity, immersion, and narrativity will therefore take the participation of a virtual body.

The writing of the first edition of this book benefited from a grant in 1998 from the National Endowment for the Humanities (grant FB-34667-98) and from a postdoctoral fellowship from the Cornell Society for the Humanities in 1999.

Though the book was conceived as a global project, not assembled from a collection of articles, many of the chapters include materials from previously published essays. The seed for the entire book is "Immersion versus Interactivity: Virtual Reality and Literary Theory," published in electronic form in *Postmodern Culture*, September 1994, reprinted by permission of the Johns Hopkins University Press. An expanded print form appeared in *Sub/Stance* 89, no. 2 (1999): 110–37, and is used by permission of the University of Wisconsin Press.

Chapter 1 contains passages from "Cyberspace, Virtuality, and the Text," published in *Cyberspace Textuality: Computer Technology and Literary Theory*, edited by Marie-Laure Ryan (Indianapolis: Indiana University Press, 1999), 78–107. Used by permission of Indiana University Press.

Chapter 5 is based on "The Text as World Versus the Text as Game: Possible Worlds Semantics and Postmodern Theory," published in *Journal of Literary Semantics* 27, no. 3 (1988): 137–63. Used by permission of Julius Groos Verlag.

Chapter 6 is based on "Between Play and Politics: Dysfunctionality in Web Art," published in *Electronic Book Review*, www.electronicbookreview.com/thread/imagenarrative/diegetic. Used by permission of the Electronic Book Review.

Chapter 7 includes materials adapted from "Peeling the Interactive Onion: Layers of User Participation in Digital Narrative Texts," published in *New Narratives: Stories and Storytelling in the Digital Age*, edited by Ruth Page and Bronwen Thomas (Lincoln: University of Nebraska Press, 2011), 35–62. Used by permission of the Board of Regents of the University of Nebraska.

Chapter 10 includes materials from two articles, "From Narrative Games to Playable Stories: Towards a Poetics of Interactive Narrative," published in *Storyworlds: A Journal of Narrative Studies* 1 (2009): 56–75 (used by permission of University of Nebraska Press), and "Interactive Narrative, Plot Types, and Interpersonal Relations," published in *Interactive Storytelling: Proceedings of 1st ICIDS 2008*, LNAI 5334, edited by Nicolas Szilas and Ulrike Spierling (Berlin: Springer Verlag, 2008), 6–13.

Interludes that appeared with chapters 2–5 and 10 in the first edition can be found at https://jhupbooks.press.jhu.edu/content/narrative-virtual-reality-2. The interludes are titled as follows:

for chapter 2: *Virtual Realities of the Mind: Baudelaire, Huysmans, Coover*
for chapter 3: *The Discipline of Immersion: Ignatius of Loyola*
for chapter 4: *Virtual Narration as Allegory of Immersion*
for chapter 5: *The Game-Reader and the World-Reader:* If On a Winter's Night a Traveler
for chapter 8: *Adventures in Hyper4text: Michael Joyce's* Twelve Blue
for chapter 10: *Dream of the Interactive Immersive Book: Neal Stephenson's* The Diamond Age

Introduction

Few of us have actually donned an HMD (head-mounted display) and DGs (data gloves), and none have entered the digital wonderland that was dangled before our eyes by the early developers of virtual reality (VR): a computer-generated, three-dimensional landscape in which we would experience an expansion of our physical and sensory powers; leave our bodies and see ourselves from the outside; adopt new identities; apprehend immaterial objects through many senses, including touch; be able to modify the environment through either verbal commands or physical gestures; and see creative thoughts instantly realized without going through the process of having them physically materialized.

When the idea of VR was introduced, with great fanfare, to the public in the early to mid-1990s, there was hardly anybody who did not have a passionate opinion about the technology: VR will someday replace reality; VR will never replace reality; VR challenges the concept of reality; VR will enable us to rediscover and explore reality; VR is a safe substitute for drugs and sex; VR is pleasure without risk and therefore immoral; VR will enhance the mind, leading us to new powers; VR is addictive and will enslave us; VR is the art of the twenty-first century, as cinema was for the twentieth; VR is pure hype and ten years from now will be no more than a footnote in the history of culture and technology. Indeed, as I revise this book in 2015, VR has mostly disappeared from the headlines, replaced in the collective imagination by applications of digital technology that affect our lives much more directly, such as mobile computing, social networking, and augmented reality. It is only in the form of virtual online worlds, which are technologically a far cry from what the pioneers of VR envisioned in the last two decades of the twentieth century, that virtual reality continues to fascinate us.

We may still have to wait a few years to see whether the promises and threats initially associated with VR will materialize. But we don't have to wait for their realization to explore the perspectives that the idea of VR opens on representation. Approaching VR as a semiotic phenomenon, I propose in this book to rethink textuality, mimesis, narrativity, literary theory, and the cognitive processing of texts in the light of the new modes of artistic world construction that have been made possible by recent developments in electronic technology.

VR has been defined as an "interactive, immersive experience generated by a computer" (Pimentel and Teixeira, *Virtual Reality,* 11). As a literary theorist, I am primarily interested in the two dimensions of the VR experience as a novel way to describe the types of reader response that may be elicited by a literary text of either the print or the electronic variety. I propose therefore to transfer the two concepts of immersion and interactivity from the technological to the literary domain and to develop them into the cornerstones of a phenomenology of reading, or, more broadly, of experiencing art. In the course of this investigation, we visit both traditional literary texts and the new genres made possible by the digital revolution of the past two decades, such as hypertext, more recent forms of electronic literature, video games, interactive drama, and electronic installation art. My purpose is twofold: to revisit print literature, specifically the narrative kind, in terms of the concepts popularized by digital culture, and, conversely, to explore the fate of traditional narrative patterns in digital culture.

The history of Western art has seen the rise and fall of immersive ideals and their displacement, in the twentieth century, by an aesthetics of play and self-reflexivity that eventually produced the ideal of active participation of the appreciator—reader, spectator, user—in the production of the text. This scenario affects both visual and literary art, though the immersive wave peaked earlier in painting than in literature.

In pre-Renaissance times, painting was more a symbolic representation of the spiritual essence of things than an attempt to convey the illusion of their presence. Its semiotic mode was signification rather than simulation. More attentive to what Margaret Wertheim (*Pearly Gates,* 87) calls "the inner eye of the soul" than to the "physical eye of the body," medieval artists painted objects as they believed them to be, not as they appeared to easily deceived senses. (The same can be said of children's drawings that represent the sky as a thin line at the top of the page rather than as a background behind figures.) Pictorial space was a strictly two-dimensional surface from which the body of the spectator was excluded, since bodies are three-dimensional objects.

All this changed when the discovery of the laws of perspective allowed the projection of a three-dimensional space onto a two-dimensional surface. This projection opens up a depth that assigns spatial coordinates—the center of projection, or physical point of view—to the body of the spectator. Perspective painting immerses a virtual body in an environment that stretches in imagination far beyond the confines of the canvas. From its spatial point of view, the embodied gaze of the spectator experiences the depicted objects as virtually present, though the flat surface of the painting erects an invisible wall that prevents physical interaction. This strictly visual immersion reached its high point in the incredible trompe l'oeil effects of the Baroque age. The frescoes of Baroque churches blur the distinction between physical and pictorial space by turning the latter into a continuation of the former.

The illusion of a penetrable space received a first challenge when impressionism disoriented the eye with visible brushstrokes that directed attention to the surface of the canvas and with shimmering light effects that blurred the contours of objects. Though impressionistic space is still three-dimensional, it opens itself to virtual bodies only after the mind completes a complex process of interpretation and construction of sensory data. For the spectator who has assimilated the lesson of impressionism, visual space can no longer be taken for granted.

In the early twentieth century, pictorial space either folded down into a play of abstract shapes and colors on a canvas that openly displayed its two-dimensionality, or it exploded into the multiple perspectives of cubist experiments. Whereas the return to flat representation expelled the body from pictorial space, the cubist approach shattered the physical integrity of both space and the body by forcing the spectator to occupy several points of view at the same time. If abstract and cubist paintings lure the spectator into a game of the imagination, this game is no longer the projection of a virtual body in a virtual space but the purely mental activity of grouping shapes and colors into meaningful configurations. As art became more and more conceptual, the eye of the mind triumphed once again over the eye of the body.

But the appeal of a pictorial space imaginatively open to the body is hard to kill off, and in the second third of the twentieth century, immersive ideals made a notorious comeback with the sharply delineated dreamscapes of surrealism. The art scene is now split between conceptual schools that engage the mind, hyperrealistic images that insist on the presence of objects to the embodied eye, and three-dimensional installation art in which the actual body is placed in an intellectually challenging environment. By letting the user walk around the display and occasionally take physical action to activate data, installation art offers a prefiguration of the combination of immersion and interactivity that forms the ideal of VR technology.

In the literary domain, no less than in the visual arts, the rise and fall of immersive ideals are tied to the fortunes of an aesthetics of illusion, which implies transparency of the medium. The narrative style of the eighteenth century maintained an ambiguous stance toward immersion: on one hand, it cultivated illusionist effects by simulating nonfictional narrative modes (memoirs, letters, autobiographies); on the other hand, it held immersion in check through a playful, intrusive narrative style that directed attention back and forth from the story told to the storytelling act. The visibility of language acted as a barrier that prevented readers from losing themselves in the storyworld.

The aesthetics of the nineteenth-century novel tipped this balance in favor of the storyworld. Through techniques that are examined in greater detail in chapter 4 of this book, high realism effaced the narrator and the narrative act, penetrated the mind of characters, transported the reader into a virtual body located on the scene of the action, and turned her into the direct witness of events, both mental and physical, that seemed to be telling themselves. Readers not only developed strong emotional ties to the characters, they were held in constant suspense by the development of the plot. The immersive quality of nineteenth-century narrative techniques appealed to such a wide segment of the public that there was no sharp distinction between "popular" and "high" literature: wide strata of society wept for Little Nell or waited anxiously for the next installment of Dickens's serial novels.

The rest of the story has been told many times: how literature, cross-fertilized with the New Criticism, structuralism, and deconstruction, took a "linguistic turn" in the mid-twentieth century, privileged form over content, and emphasized spatial relations between words, puns, intertextual allusion, parody, and self-referentiality. The novel subverted plot and character, experimented with open structures and permutations, turned into increasingly cerebral wordplay, or became indistinguishable from lyrical prose. This evolution split literature into an intellectual avant-garde committed to the new aesthetics and a popular branch that remained faithful to the immersive ideals and narrative techniques of the nineteenth century. (Ironically, the high branch turned out to be heavily dependent on the resources of the low branch for its game of parody.) As happened in the visual arts, immersion was brought down by a playful attitude toward the medium, which meant in this case the exploitation of such features as the phonic substance of words, their graphic appearance, and the clusters of related or unrelated senses that make up their semantic value field. In this carnivalesque conception of language, meaning was no longer the stable image of a world in which the reader projects a virtual alter ego, nor even the dynamic simulation of a world in time, but the sparks generated by associative chains that connect the particles of a textual and intertextual field of energies into ever-changing configurations. Meaning came to be described as unstable, decentered, multiple, fluid, emergent—all concepts that have become hallmarks of postmodern thought.

Though this game of signification needs nothing more than the encounter between the words on the page and the reader's imagination to be activated, it is easy to see how the feature of interactivity conferred upon the text by electronic technology came to be regarded as the fulfillment of the postmodern conception of meaning. Interactivity takes the ideal of an endlessly self-renewable text from the level of the signified to the level of the signifier. In hypertext, the prototypical form of interactive textuality (though by no means the most interactive), the reader determines the unfolding of the text by clicking on certain areas, the so-called hyperlinks, that bring to the screen other segments of text. Since every segment contains several such hyperlinks, every reading produces a different text, if by *text* one understands a particular set and sequence of signs scanned by the reader's eye. Whereas the reader of a standard print text constructs personalized interpretations out of an invariant semiotic base, the reader of an interactive text thus participates in the construction of the text as a visible display of signs. Although in hypertext this process is restricted to a choice among a limited number of well-charted alternatives—namely, the branching possibilities designed by the author—this relative freedom has been hailed as an allegory of the vastly more creative and less constrained activity of reading as meaning formation.

While interactivity has been celebrated by the early theorists of hypertext as a liberation from authorial power and from the linearity of language, the concept of immersion has suffered a vastly different fate, at least until the late 1990s. If we believe some of the most celebrated works of world literature, losing oneself in a book, or in any kind of virtual reality, is a hazard for the health of the mind. Warnings about the ravages of immersion appear as early as in the first great novel of European literature. "In short," writes Cervantes in *Don Quixote,* "he so immersed himself in those romances that he spent whole days and nights over his books; and thus with little sleeping and much reading, his brains dried up to such a degree that he lost the use of his reason" (58). The situation does not seem to be better in the virtual realities of the electronic kind: we hear tales of people suffering from AWS (alternate world syndrome), a loss of balance, feeling of sickness, and general "body amnesia" (Heim, *Virtual Realism,* 52) when they leave VR systems, of video games addicts who cannot adapt to ROL (Sherry Turkle's acronym for "the rest of life"), or of children who experience emotional trauma when they inadvertently let their virtual pets die.

The major objection against immersion is the alleged incompatibility of the experience with the exercise of critical faculties. The semiotic blindness caused by immersion is illustrated by an anecdote involving the eighteenth-century French philosopher Diderot. As Wallace Martin reports, "He tells us how he began reading *Clarissa* several times in order to learn something about Richardson's techniques, but never succeeded in doing so because he became personally involved in the work, thus losing his critical consciousness" (*Recent Theories,* 58). According to Jay David Bolter, the impairment of critical consciousness is the trademark of both literary and VR immersion: "But is it obvious that virtual reality cannot in itself sustain intellectual or cultural development. . . . The problem is that virtual reality, at least as it is now envisioned, is a medium of percepts rather than signs. It is virtual television" (*Writing Space,* 230). And further: "Losing oneself in a fictional world is the goal of the naive reader or one who reads as entertainment. It is particularly a feature of genre fiction, such as romance or science fiction" (155).

Critical recognition of immersion as a valuable aesthetic experience has not been helped by its resistance to theorization. What can be said about immersion in a textual world except that it takes place? The self-explanatory character of the concept is easily interpreted as evidence that immersion promotes a passive attitude in the reader, similar to the entrapment of tourists in the self-enclosed virtual realities of theme parks or vacation resorts. The association of the experience with popular culture only reinforces this charge. Because of its reliance on stereotypes, devotees of high-brow and avant-garde literature accuse popular culture of turning the reading experience into something like taking a dip in a Jacuzzi: it is easy to get in, but you cannot stay in very long, and you feel tired once you get out.

But this does not mean that immersive pleasure is in essence a lowbrow, escapist gratification. At its best, immersion can be an adventurous and invigorating experience comparable to taking a swim in a cool ocean with powerful surf. The environment at first appears hostile, you enter it reluctantly, but once you get wet and entrust your body to the waves, you never want to leave. And when you finally do, you feel refreshed and full of energy. As for the allegedly passive character of the experience, we need only be reminded of the complex mental activity that goes into the production of a vivid mental picture of a textual world. Since language does not offer input to the senses (except for the aural or visual properties of words, which are normally not related to their meaning), all sensory data must be simulated by the imagination. In "The Circular Ruins," Jorge Luis Borges writes of the protagonist, who is trying to create a human being by the sheer power of his imagination, "He wanted to dream a man: he wanted to dream him with minute integrity and insert him into reality" (*Ficciones,* 114). Similarly, we must dream up textual worlds with "minute integrity" to conjure the intense experience of presence that inserts them into imaginative reality. Is this the trademark of a passive reader?

Since the mid-1990s, fortunately, the tide has been turning on the concept of immersion. Interest has come from New Media theorists, such as Janet Murray, who regards the experience as essential to the aesthetics of the electronic medium, and from game designers, who depend on immersion for commercial success. Among literary scholars, Jean-Marie Schaeffer makes immersion into the cornerstone of a theory of fictionality, and Werner Wolf regards it as an essential component of what he terms "aesthetic illusion," an experience that can be provided by both literature and the visual arts. Meanwhile, psychologist Keith Oatley developed a concept of mental simulation that could be seen as responsible for immersion. This increased focus on immersion can be attributed to several sources: the waning of the influence of "textualist" schools on literary criticism (i.e., schools that describe the literary text as a system of signs held together by horizontal relations between signifiers, neglecting vertical relations to a world), the rehabilitation of popular culture, embodied theories of cognition (to be immersed is to take a virtual body into an imaginary world), the rise of interest in emotions (empathy with characters is a major source of immersion), and advances in cognitive science and brain imaging that reveal the neurophysiological basis of mental simulation (the mirror neurons system). Clearly, there is now much more to say about immersion than that it just takes place.

The organization of this book grew out of the very definition that inspired the whole project: "virtual reality is an immersive, interactive experience generated by a computer." We begin by visiting the virtual as philosophical concept; move on to VR as technology; explore its two components, immersion and interactivity; and conclude by considering what is (for me) the ultimate goal of art: the synthesis of immersion and interactivity. This book, then, is as much about virtual literature—literature that could be—as about existing texts.

Judging by their current popularity in both theory and advertising language, the terms *virtual* and *virtuality* exert a powerful magnetism on the contemporary imagination, but as is always the case when a word catches the fancy of the general public, the meaning tends to dissolve in proportion to the frequency of use. In its everyday usage the word *virtual* is ambiguous, used in the sense of (1) "imaginary" as well as (2) "depending on computers." (A third sense derived from scholastic philosophy, the virtual as potential, does not seem as influential on the popular usage.) When we speak of "virtual pets," we mean the computer image of corporeally nonexistent animal companions, but when we speak of "virtual technologies," we certainly do not mean something that does not exist, or we would not spend hundreds of dollars for computer software. Virtual technologies fabricate objects that are virtual in sense 1 but are themselves virtual in sense 2. When N. Katherine Hayles characterizes the condition of contemporary mankind as "virtual" and further defines this condition as "the cultural perception that material objects are interpenetrated by information patterns" ("Condition of Virtuality," 69), she makes a culturally well accepted, but philosophically less evident, association: Why should information be regarded as virtual or at least as meaningfully connected with virtuality? Is it because information enables us to build "virtual realities"—digital images that offer simulacra of physically habitable environments? Is it because informational patterns contain *in potentia* new forms of life (as in biological engineering), new forms of art, and, for the dreamers of the coupling of man and machine, new forms of humanity? Is it because information lives principally these days in the silicon memory of computers, invisible and seemingly inexistent until the user summons it to the screen?

I have suggested here three distinct senses of *virtual*: an optical one (the virtual as illusion), a scholastic one (the virtual as potentiality), and an informal technological one (the virtual as the computer mediated). All three are involved in VR: the technological because VR is made of digital data generated by a computer; the optical because the immersive dimension of the VR experience depends on the reading of the virtual world as autonomous reality, a reading facilitated by the illusionist quality of the display; and the scholastic because, as interactive system, VR offers to the user a matrix of actualizable possibilities. In the first chapter of this book, I explore the optical and the scholastic interpretation of the virtual by relating them to the work of two prominent French theorists—Jean Baudrillard for the virtual as illusion and Pierre Lévy for the virtual as potentiality. I dwell on these two versions of the virtual not only for the sake of their involvement with VR technology but also because each of them presents important implications for literary theory and the phenomenology of reading.

In the second chapter, I turn to VR proper. Though the current state of the technology falls way short of the expectations raised at the time of its first introduction to the general public, the "myth" matters as much as the technological reality for a project that uses VR as metaphor, and I therefore move back and forth between the exalted vision of the early prophets and the more sober descriptions of the technical literature. Immersion in a virtual world is discussed from both a technological and a phenomenological point of view. Whereas the technological approach asks what features of digital systems produce an immersive experience, the phenomenological discussion analyzes the sense of "presence" through which the user feels corporeally connected to the virtual world. I look for answers to this second question in the writings of a philosopher acutely aware of the embodied nature of perception, Maurice Merleau-Ponty. If these concerns seem to emphasize immersion to the detriment of interactivity, it is not because VR subordinates one to the other—it may or it may not, depending on its ultimate purpose—but because immersion is by far the more problematic concept. We all know instinctively what interactivity consists of in a computer program—submitting input and receiving output—but it is much harder to tell what it means to feel immersed in a virtual world and how digital technology and interface design can promote this experience.

The phenomenological idea of consciousness as a sense of being-in-the-world —or in this case, in a simulated world—is at the core of the theory and poetics of immersion presented in the second part of the book. The term *immersion* has become so popular in contemporary culture that people tend to use it to describe any kind of intensely pleasurable artistic experience or any absorbing activity. In this usage, we can be immersed in a crossword puzzle as well as in a novel, in the writing of a computer program as well as in playing the violin. Here, however, I would like to single out and describe a specific type of immersion, one that presupposes an imaginative relationship to a *world projected by a text*—an intuitive concept to be refined in chapter 3. In the phenomenology of reading, immersion is the experience through which a fictional world acquires the presence of an autonomous, language-independent reality populated with live human beings.

For a text to be immersive, then, it must create a space to which the reader, spectator, or user can relate, and it must populate this space with individuated objects. It must, in other words, construct the setting for a potential narrative action, even though it may lack the temporal extension to develop this action into a plot. This fundamentally *mimetic* concept of immersion remains faithful to the VR experience, since the purpose of VR technology is to connect the user to a simulated reality. It applies to novels, movies, drama, representational paintings, and those computer games that cast the user in the role of a character in a story, but it does not apply to philosophical works, music, and purely abstract games such as bridge, chess, and Tetris, no matter how absorbing these experiences can be.

Immersion may not have been particularly popular with the "textual" brands of literary theory, but this does not mean that the experience was totally ignored when the critical scene was dominated by these theories. Chapter 3 discusses the work of some scholars working on the outskirts of literary studies—in cognitive psychology, empirical approaches to literature, or analytic philosophy—who have addressed the issue that I call immersion, though they have done so under a variety of other names. This work includes Victor Nell's analysis of the psychological state of being "lost in a book"; Richard Gerrig's concept of transportation; the possible-worlds approach to the semantics of fictionality and its description of the phenomenology of reading fiction as an imaginative "recentering" of the universe of possibilities around a new actual world; and Kendall Walton's theory of fiction as a game of make-believe. The chapter concludes with recent attempts to explain what it means to be immersed through the concept of mental simulation.

Whether textual worlds function as imaginary counterparts or as models of the real world, the reader mentally constructs them as environments that stretch in space, exist in time, and serve as habitats for a population of animate agents. These three dimensions correspond to what have long been recognized as the three basic components of narrative grammar: setting, plot, and characters. The "poetics" proposed in chapter 4 associates these narrative elements with three distinct types of immersion—spatial, temporal, and emotional—and analyzes the narrative devices that favor each of them. In my discussion of temporal and emotional immersion, I seek explanations for two closely related immersive paradoxes that have generated lively debate among philosophers and cognitive psychologists for a number of years: how readers can experience suspense the second or third time they read a text, even though they know how it ends, and how the fate of fictional characters can generate emotional reactions with physical symptoms, such as crying, even though readers know full well that these characters never existed.

Chapter 5 examines the change of metaphor that marked the transition from immersion to interactivity as artistic ideals. Whereas the aesthetics of immersion implicitly associates the text with a "world" that serves as environment for a virtual body, the aesthetics of interactivity presents the text as a game, language as a plaything, and the reader as the player. The idea of verbal art as a game with language is admittedly not a recent invention; ancient literatures and folklore are full of intricate word games, and the novel of the eighteenth century engaged in very self-conscious games of narration. But it was only in the middle of the twentieth century, after the concept of "game" rose to prominence as a philosophical and sociological issue and began infiltrating many other disciplines, that literary authors developed the metaphor into an aesthetic program. The concept of "game" covers, however, a wide variety of activities, and literary critics too often use it in a generic sense. Chapter 5 narrows down the metaphor by exploring what kind of games and what specific features pertaining to these games provide meaningful analogies with the literary domain.

In chapter 6, I look at digital texts that reject world aesthetics in favor of game aesthetics, thereby ostentatiously preventing immersion. The result, I argue, is deliberate dysfunctionality, a phenomenon that can be observed in many different domains: the domain of code, the domain of applications (such as e-mail or graphic programs), the domain of games, the domain of interface, and more fundamentally, the domain of language. By emphasizing play at the expense of mimesis, these texts enact a conceptual aesthetics that subverts the mode of signification embodied by language. Can these texts still be read, or are they meant to be simply looked at?

The remaining chapters of the book are devoted to the problem of combining world aesthetics with game aesthetics by introducing interactive elements into storyworlds, thereby bringing them closer to life experience, since it is by acting upon our environment that we relate to the real world. Chapter 7 lays down the foundations of a theory of interactivity by analyzing it according to several criteria. A first taxonomy is based on the cross-classification of two dichotomies: the relation of the interactor to the textual world, which can be either internal or external, and the status of her actions, which can be ontological (creating changes in the world) or merely exploratory. Then I examine various types of textual architecture that support interactivity. Comparing texts to onions with many layers of skin, the last section classifies interactivity according to the narratological level that it affects, from the outer level of the presentation of data to the inner level of the plot.

In chapter 8 I turn to hypertext, the genre of digital fiction that was hailed in the nineties as the future of narrative. With the hindsight gained through the twenty years that have elapsed since the first literary hypertexts became available, I dissect the critical discourse that celebrated hypertext as the embodiment of what I have called the "textualist" conception of literary art: an endlessly self-renewable artifact whose signification cannot be exhausted. Yet Roland Barthes, whose vision of a plural text has deeply influenced hypertext theory, viewed plurality as incompatible with narrativity. Was he right? Can narrativity, a reasonably solid structure, be maintained in a fluid environment in which sequence cannot be controlled by the author? To answer this question, I examine various ways of conceptualizing the hypertext experience: as exploring a territory, as playing with a kaleidoscope, as selecting goods from a shelf, as solving a jigsaw puzzle, as trying to get out of a maze, as visiting possible worlds, and so on. Which one fits the best depends on the text's underlying structure and on the function of links, but none of these conceptualizations describes the genre as a whole nor exhausts the possibilities of individual texts.

Even when narrative coherence is maintained, immersion remains an elusive experience in interactive texts. In the last two chapters, I argue that the marriage of immersion and interactivity requires the imagined or physical presence of the appreciator's body in the virtual world—a condition easily satisfied in a VR system but problematic in hypertext because every time the reader is asked to make a choice she assumes an external perspective on the worlds of the textual universe. In VR we act within a world and experience it from the inside, but in interactive texts of the selective variety we choose a world, more or less blindly, out of many alternatives, and we are not imaginatively committed to any one of them, because the interest of branching texts lies in the multiplicity of paths, not in any particular development.

As chapter 9 shows, VR is not the only environment that offers an experience both immersive and interactive: children's and adults' games of make-believe, fairs and amusement parks, ritual, Baroque art and architecture, and certain types of stage design in the theater propose an active participation of either an actual or virtual body in a reality created by the imagination. The study of these experiences should therefore provide valuable guidelines for the design of electronic texts.

In chapter 10, I leave the domain of actually existing texts for the domain of possible designs. I am concerned with ways to deal with what Ruth Aylett and Sandy Louchard have called the interactive paradox: providing a top-down, author-controlled narrative design while giving the interactor a sense of freedom of action. After distinguishing two types of interactive narrative—narrative games and playable stories—I turn to four kinds of plots found in traditional literature (the epic, the dramatic, the epistemic, and the soap opera) in order to assess their ability to accommodate an active user. Since combining narrativity with interactivity would not be a valid goal if it happened at the cost of immersivity, I conclude the chapter with an assessment of the potential of interactive narrative to satisfy the three types of immersion that I have defined in chapter 4: spatial, temporal, and emotional.

By proposing to read VR as a metaphor for total art, I do not mean to suggest that the types of art or entertainment discussed in these last two chapters are superior to the mostly immersive forms of part II or the mostly interactive ones of part III. If aesthetic value could be judged by numerical coefficients, as in certain "artistic" sports such as equestrian dressage or figure skating, a text that scored ten on immersion and one on interactivity—a good realistic novel—would place higher than a text that scored three for each criterion. Whether or not future VR installations will be able to offer more than mediocrity on both counts, however, we can still use the *idea* of VR as a metaphor for the fullest artistic experience, since in the Platonic realm of ideas VR scores a double ten.

But why should the synthesis of immersion and interactivity matter so much for aesthetic philosophy? In its literal sense, immersion is a corporeal experience, and as I have hinted, it takes the projection of a virtual body—or even better, the participation of the actual one—to feel integrated in an art-world. On the other hand, if interactivity is conceived as the appreciator's engagement in a play of signification that takes place on the level of signs rather than things and of words rather than worlds, it is a purely cerebral involvement with the text that downplays emotions, curiosity about what will happen next, and the resonance of the text with personal memories of places and people. On the shiny surface of signs—the signifier—there is no room for bodies of either the actual or the virtual variety. It is only when interactivity is conceived as action-upon-a-world that it involves mind and body. What is at stake in the synthesis of immersion and interactivity is therefore nothing less than the participation of the whole of the individual in the artistic experience.

I: Virtuality

ONE

The Two (and Thousand) Faces of the Virtual

I dwell in Possibility
A fairer House than Prose,
More numerous of Windows,
Superior for Doors.

Of Chambers, as the Cedars—
Impregnable of eye;
And for an everlasting Roof
The Gambrels of the Sky.

Of Visitors—the fairest—
For Occupation—This—
The spreading wide my narrow
Hands To gather Paradise—

—*Emily Dickinson*

In the popular imagination of the last decade of the twentieth century, the word *virtual* triggers almost automatically the thought of computers and digital technology. This association was built in several steps, though the early ones have largely fallen into oblivion. Nowadays we label virtual everything we experience or meet in "cyberspace," the imaginary place where computers take us when we log on to the Internet; we can meet virtual friends, have virtual sex, study at virtual universities, and go on virtual tours of virtual cities. Before the Internet forced itself, almost overnight, into our daily lives, the virtuality of digital technology was associated with the concept of virtual reality (VR), introduced to the public in the late 1980s. Computers were credited with the power to create artificial worlds, and although the Internet is a far cry from the three-dimensional, multisensory, immersive, and interactive environments envisioned by the promoters of VR, we have projected onto cyberspace the dreams that the VR industry awakened but largely failed to deliver.

Earlier in the history of the semantic liaison of *virtual* and *reality, virtual* was a technical term of computer architecture that expressed the discrepancy between the physical machine and the machine with which users and high-level programmers think they are communicating. Computer programs are written in a quasi-human language made up of a large number of powerful modules and commands, but the actual processor can understand only a small number of instructions coded in zeros and ones. It takes a translator, known as a compiler or an interpreter, to turn the instructions typed by the user into executable code. In the same vein, the term *virtual* was applied to memory to refer to a type of storage, such as a floppy disk, that is not physically part of the computer's active memory but whose contents can easily be transferred back and forth to the brain of the machine, so that from the point of view of the user this storage behaves as if it were an integral and permanent part of computer memory.

Yet another virtual feature of computers resides in their versatility. As a machine, a computer has no intrinsic function. Through its software, however, it can simulate a number of existing devices and human activities, thus becoming a virtual calculator, typewriter, record player, storyteller, babysitter, teacher, bookkeeper, or adviser on various matters. It can even become, as VR suggests, a virtual world and living space. The software industry exploited these technical uses—of which there are many others—by metonymically promoting its products as "virtual technologies." For the general public, the narrow technical meaning meant nothing, but the label *virtual* became a powerful metaphor for the accelerating flight of technology into the unknown. The term gave an almost science-fictional aura to the products of a culture that had to be hatching something fundamentally new, since it was approaching the mythical landmark of the turn of the millennium.

Let us backtrack even further, in this hopeless but tempting search for pure and original meaning, by asking what is virtual about artificial worlds and pseudo-memory and versatile machines. Etymology tells us that *virtual* comes from the Latin *virtus* (strength, manliness, virtue), which gave to scholastic Latin the philosophical concept of *virtus* as force or power. (This sense survives today in the expression "by virtue of.") In scholastic Latin, *virtualis* designates the potential, "what is in the power [*virtus*] of the force." The classic example of virtuality derived from Aristotle's distinction between potential and actual existence (*in potentia* vs. *in actu*) is the presence of the oak in the acorn. In scholastic philosophy, "actual" and "virtual" exist in a dialectical relation rather than in one of radical opposition: the virtual is not that which is deprived of existence but that which possesses the potential, or force, of developing into actual existence. Later uses of the term, beginning in the eighteenth and nineteenth centuries, turn this dialectical relation to actual into a binary opposition to real: the virtual becomes the fictive and the nonexistent. This sense is activated in the optical use of the term.

According to Webster's dictionary, a virtual image, such as a reflection in a mirror, is one made of virtual foci, that is, of points "from which divergent rays of light seem to emanate but do not actually do so." Exploiting the idea of fake and illusion inherent to the mirror image, modern usage associates the virtual with that which *passes as* something other than what it is. This passing involves an element of illegitimacy, dishonesty, or deficiency with respect to the real. A virtual dictator may be "as good"—or in this case as bad—as a real dictator, but he remains inferior to a "legitimate" one, to use an oxymoron, because he is not officially recognized as such. (He could, in principle, be indicted for abuse of power.) Yet the deficiency of the virtual with respect to the real may be so small that for all practical purposes the virtual becomes the real.

As we see from these lexical definitions, the meaning of *virtual* stretches along an axis delimited by two poles. At one end is the optical sense, which carries the negative connotations of double and illusion (two ideas combined in the theme of the treacherous image); at the other is the scholastic sense, which suggests productivity, openness, and diversity. Somewhere in the middle are the late-twentieth-century associations of the virtual with computer technologies. For convenience's sake I will refer to one pole of the virtual as fake and the other as potential. (See table 1 for a list of the connotations I collected in the course of my various readings on virtuality.) Both of these interpretations have found influential and eloquent spokesmen in recent French theory: Jean Baudrillard for the virtual as fake, Pierre Lévy for the virtual as potential.

Baudrillard and the Virtual as Fake

The philosophy of Baudrillard presents itself as a meditation on the status of the image in a society addicted to "the duplication of the real by means of technology" (Poster, "Theorizing," 42). Once, the power to automatically capture and duplicate the world was the sole privilege of the mirror; now this power has been emulated by technological media—photography, movies, audio recordings, television, and computers—and the world is being filled by representations that share the virtuality of the specular image. The general tone and content of Baudrillard's meditation on this state of affairs are given by the epigraph to his most famous essay, "The Precession of Simulacra," a quotation attributed to Ecclesiastes that is nowhere to be found in the Bible (hence, evidently, the lack of reference to a verse number). True to its message and subject matter, the essay thus opens with a simulacrum: "The simulacrum is never what hides the truth—it is truth that hides the fact that there is none. The simulacrum is true" (1). A simulacrum, for Baudrillard, is

Table 1. The meaning of virtual

Actual	Virtual	Actual	Virtual
enacted	potential	presence	absence
factual	counterfactual	face-to-face	mediated
accomplished	possible	mechanical, printed	electronic
closed	open	matter	information
material	mental	space	cyberspace
concrete	abstract	that which counts	that which does not count
complete	incomplete	being	presenting
determinate	indeterminate	identity	passing as, role-playing
corporeal	spectral	serious behavior	make-believe
bound body or object	aura	lived experience	fantasy and dreams
kernel	irradiation	fact	fiction
temporally located	temporally free-floating	essence	appearance
inscribed in space	deterritorialized	authenticity	fake, simulation
singular	plural	truth	illusion, falsity
manufactured object	blueprint, code	original	copy, double
present	past and future	represented, referent	image
here	there	virtual	real (theme of the virtual and real exchanging places)
solid, tangible	evanescent, nontangible	. . .	real (theme of the disappearance of the real)
figure	ground		
visible	latent		

Note: At the top left, the scholastic opposition of virtual to actual. At the bottom right, the optical opposition of virtual to real. In between (assuming the table runs as a single column), the popular association of virtual with digital technology.

not the dynamic image of an active process, as are computer simulations, but a mechanically produced and therefore passively obtained duplication whose only function is to *pass as that which it is not*: "To simulate is to feign to have what one doesn't have" (3). Baudrillard envisions contemporary culture as a fatal attraction toward simulacra. This "will to virtuality," to borrow Arthur Kroker and Michael

Weinstein's evocative term (*Data Trash,* chap. 3), precludes any dialectical relation and back-and-forth movement between the real and its image. Once we break the second commandment, "Thou shalt not make images," we are caught in the gravitational pull of the fake, and the substance of the real is sucked out by the virtual, for as Baudrillard writes in *The Perfect Crime,* "There is no place for both the world and its doubles" (34). In the absence of any Other, the virtual takes the place of the real and becomes the hyperreal. In Baudrillard's grandiose evolutionary scheme, we have reached stage four in the evolution of the image:

1. It is the reflection of a profound reality.
2. It masks and denatures a profound reality.
3. It masks the absence of a profound reality.
4. It has no relation to any reality whatsoever: it is its own pure simulacrum. ("Precession," 6, numbering mine)

Does the seemingly inevitable historical evolution from stage one to stage four represent a fall into inauthenticity, an abdication of representational responsibility, and a cynical betrayal of the Real, or, on the contrary, a gradual discovery of the True Nature of the image? Has the culture of illusion committed a "perfect crime" that killed reality without leaving any traces, as Baudrillard suggests in the later book by that title, or has it definitively slain the illusion of the real and reached the ultimate semiotic wisdom? Oscillating between the roles of modern-day Ecclesiastes and solemn theorist of semiotic nihilism—and obviously enjoying himself in both roles—Baudrillard is careful to maintain an ambiguous stance.

The word *virtual* itself is absent from "The Precession of Simulacra," an essay written in the late 1970s, when the principal channel of "the image" and the main threat to the real was television. But when computer technology began to impose the notion of virtuality, in the late 1980s, Baudrillard suddenly discovered a new culprit for modern society's "crime against reality." It is as if technology had caught up with the theory and turned it into prophecy by delivering its missing referent. As Mark Poster writes, "Baudrillard's writing begins to be sprinkled with the terms 'virtual' and 'virtual reality' as early as 1991. But he uses these terms interchangeably with 'simulation,' and without designating anything different from the earlier usage" ("Theorizing," 45). In *The Perfect Crime* (1996), virtual reality is treated not as just another way to produce simulacra but as the ultimate triumph of the simulacrum.

> With the Virtual, we enter not only upon the era of the liquidation of the Real and Referential, but that of the extermination of the Other.

> It is the equivalent of an ethnic cleansing which would not just affect particular populations but unrelentingly pursue all forms of otherness.
> The otherness . . .
> Of the world—dispelled by Virtual Reality. (*Perfect Crime*, 109)

According to Baudrillard, we don't live in a world where there is something called VR technology; rather, we are immersed in this technology; we live and breathe virtual reality. All the concepts and buzzwords associated with VR provide easy fuel for Baudrillard's insatiable theoretical machine. Consider the following passages from "Aesthetic Illusion and Virtual Reality."[1]

On the transparency of the medium, one of the acknowledged goals of VR developers:

> And if the level of reality decreases from day to day, it's because the medium itself has passed into life, and become a common ritual of transparency. It is the same for the virtual: all this digital, numerical and electronic equipment is only the epiphenomenon of the virtualization of human beings in their core. (*Art and Artefact*, 20)

By the same logic that denies a place for both the world and its doubles, there is no place in the mind for both life and the lifelikeness of transparent media. Our fascination with the latter turns us into "virtual beings" through a reasoning that skips several intermediary steps in one powerful leap: (1) VR technology (and modern media in general) aims toward transparency; (2) transparency allows immersion; (3) by a metonymic transfer, immersion in a virtual world leads to a virtualization of the experiencer. One must assume that this virtualization involves a loss of humanity, as we offer ourselves as data and as servants to the machine.

On the project of creating three-dimensional environments with which the user can interact:

> For example some museums, following a sort of Disneyland processing, try to put people not so much in front of the painting—which is not interactive enough and even suspect as pure spectacular consumption—but into the painting. Insinuated audiovisually into the virtual reality of the *Déjeuner sur l'herbe* [by Renoir], people will enjoy it in real time, feeling and tasting the whole Impressionist context, and eventually interacting with the picture. The masses usually prefer passive roles and avoid representation. This must change, and they must be made interactive partners. It is not a question of free speaking or free acting—just break their resistance and destroy their immunities. (22)

In this passage Baudrillard's a priori commitment to the idea that we are prisoners of our own technologies of representation allows only one interpretation of interactivity: it is a simulacrum of activity that conceals the fundamental passivity of the user, just as the world outside prisons is for Baudrillard a simulacrum of freedom that conceals the fundamentally carceral nature of society ("Precession," 12).

On the digital coding of information:

> Now what exactly is at stake in this hegemonic trend towards virtuality? What is the idea of the virtual? It would seem to be the radical actualization, the unconditional realization, of the world, the transformation of all our acts, of all historical events, of all material substance and energy into pure information. The ideal would be the resolution of the world by the actualization of all facts and data. (23)

If reality has become an edifice of digital information, any bug or virus can bring the end of the world. We have seen the effects of the literalization of this belief in the millenarian hysteria of Y2K cultism.

On telepresence:

> Artificial intelligence, tele-sensoriality, virtual reality and so on—all this is the end of illusion. The illusion of the world—not its analytical countdown—the wild illusion of passion, of thinking, the aesthetic illusion of the scene, the psychic and moral illusion of the other, of good and evil (of evil especially, perhaps), of true and false, the wild illusion of death, or of living at any price—all this is volatilized in psychosensorial telereality, in all these sophisticated technologies which transfer us to the virtual, to the contrary of illusion: to radical disillusion. (27)

Why is virtual reality the end of illusion? Because it posits the deliberate and cynical choice of the virtual as fake over the world, as if we faced an absolute binary choice—live in the real, or live in the virtual—and as if we were seduced by the virtual into making the wrong choice. In this black-or-white vision, once we enter the virtual worlds of modern media, they close down upon us, and there is no way back to the real. Further on, however, Baudrillard seems to switch sides, gleefully warning us that "fortunately all this is impossible" (27), as if we had invested our hopes in this dystopian vision. Because of technological limitations, VR will never deliver on its promise to provide a perfect duplicate of reality. So what is there to fear? For the numerous admirers of Baudrillard, the value of his thought is less as a description of the real—or of the place of the virtual in the real—than as a theory of the *what if.* What if VR were perfectly realized? Would we spend our entire lives inside a Disneyland of digital data? Would images become our world? How would we tell the difference between simulation and reality? If we could not do so, would this mean that simulations *had become* reality—or alternatively, that reality *was* a simulation?

By asking us to entertain hypothetical situations and dystopic possibilities, Baudrillard theorizes the triumph of the virtual as fake as something contained in the virtual as potential, but his language creates a fake all its own by hyperbolically couching the potential in the language of actuality. The real does not *threaten* to disappear in Baudrillard's text; it has already been killed *as a matter of fact.* Readers have no problems undoing the hyperbole and linking the theory to real-world tendencies rather than to terminal states of affairs: the invasion of culture by visual representations; the control of the mind by the media; the voracious appetite of modern society for images, an appetite that sometimes tempts us to kill the real in order to produce simulacra; and last but not least a cultural fascination with the hyperreal, a copy more real than the real that destroys the desire for the original. For many cultural critics who draw inspiration from Baudrillard, the real has not disappeared; it has merely exchanged places with the virtual, in the admittedly watered-down sense that time spent in the virtual seems to grow at an alarming rate in lives that ought to be, or used to be, rooted in a solid reality. Like the Golden Calf—the image that broke the second commandment at the very moment it was given—our gods are virtual when they should be real.

But Baudrillard himself has another idea of the ontological status of his discourse. If we have reached stage four in the evolution of the image, this means that his theory falls under the scope of its own pronouncement and becomes one of these simulacra that engender their own reality. In "Radical Thought," an essay from *The Perfect Crime,* Baudrillard distinguishes two kinds of thought and leaves no doubt as to where he situates his own.

> A certain form of thought is bound to the real. It starts out from the hypothesis that ideas have referents and that there is a possible ideation of reality. A comforting polarity, which is that of tailor-made dialectical and philosophical solutions. The other form of thought is eccentric to the real, a stranger to dialectics, a stranger even to critical thought. It is not even a disavowal of the concept of reality. It is illusion, power of illusion, or, in other words, a playing with reality, as seduction is a playing with desire, as metaphor is a playing with truth. (96)

The ultimate is for an idea to disappear as idea, to become a thing among things. (100)

So, for example, you put forward the idea of the simulacrum, without really believing in it, even hoping that the real will refute it (the guarantee of scientificity for Popper).

Alas, only the fanatical supporters of reality react: reality, for its part, does not seem to wish to prove you wrong. Quite the contrary, every kind of simulacrum parades around it. And reality, filching the idea, henceforth adorns itself with all the rhetoric of simulation. It is the simulacrum which ensures the continuity of the real today, the simulacrum which conceals not the truth, but the fact that there isn't any [stage 3]—this is to say, the continuity of the nothing . . .

It's terrifying to see the idea coincide with the reality. (101)

Radical thought encounters no resistance from the real, because in contrast to the "regular" brand—the kind that does not understand itself and that lives in the illusion of referentiality—it conceives its mode of operation as declarative rather than as descriptive. Like fictional discourse, it inhabits not the true-or-false but the true by say-so. In *The Perfect Crime,* as Poster observes, "Baudrillard has become virtual and knows himself to be such: he argues that his critical theory of simulation has become the principle of reality" ("Theorizing," 46). As the representation, or virtual, that becomes reality, Baudrillard's theory embodies, literally, the paradoxical idea of *virtual* reality.

For those who are prevented by an enduring sense of the presence and alterity of the real from accepting the idea that it derives from Baudrillard's discourse (should we call this sense simply "common"?), there remains fortunately the alternative of a nonradical interpretation. We live in simulacra because we live in our own mental models of reality. What I call "the world" is my perception and image of it. Therefore, what is real for me is the product of my copy-making, virtual-producing, meaning-making capability. The copies that make up my world cannot be perfect duplications, but this does not make them necessarily false, deceptive, or deprived of referent. In this interpretation, the absolutely real has not disappeared; it is, rather, as Slavoj Zižek defines it, "a surplus, a hard kernel which resists any process of modeling, simulation, or metaphorization" (*Tarrying,* 44). We know that this "other" real exists, and often we butt into it, but we do not live in it, except perhaps in some moments of thoroughly private and nearly mystical experience, because the human mind is an indefatigable fabricator of meaning, and meaning is a rational simulacrum of things. Disarming the other of its otherness by representing it and building "realities" as worlds to inhabit are one and the same thing. It is simply thinking.

Lévy and the Virtual as Potential

Becoming Virtual, the English title of Pierre Lévy's *Qu'est-ce que le virtuel,* may seem at first sight to confirm Baudrillard's most pessimistic prediction for the future of humanity. But the impression is dispelled as early as the second page of the introduction to Lévy's treatise: "The virtual, strictly defined, has little relationship to that which is false, illusory, or imaginary. The virtual is by no means the opposite of the real. On the contrary, it is a fecund and powerful mode of being that expands the process of creation, opens up the future, injects a core of meaning beneath the platitude of immediate physical presence" (16).

Lévy outlines his concept of virtuality—inspired in part by Gilles Deleuze's ideas on the topic—by opposing two conceptual pairs: one static, involving the possible and the real, and the other dynamic, linking the actual to the virtual.[2] The possible is fully formed, but it resides in limbo. Making it real is largely a matter of throwing the dice of fate. In the terminology of modal logic, this throw of the dice may be conceived as changing the modal operator that affects a proposition, without affecting the proposition itself. All it takes to turn the possibility into the actuality of a snowstorm is to delete the symbol $\Diamond$ (possibility operator) in front of the proposition "It is snowing today." The operation is fully reversible, so that the proposition p can pass from mere possibility to reality back to possibility. In contrast to the predictable realization of the possible, the mediation between the virtual and the actual is not a deterministic process but a form-giving force. The pair virtual/actual is characterized by the following features:

1. The relation of the virtual to the actual is one-to-many. There is no limit on the number of possible actualizations of a virtual entity.
2. The passage from the virtual to the actual involves transformation and is therefore irreversible. As Lévy writes, "Actualization is an *event,* in the strongest sense of the term" (171).
3. The virtual is not anchored in space and time. Actualization is the passage from a state of timelessness and deterritorialization to an existence rooted in a here and now. It is an event of contextualization.
4. The virtual is an inexhaustible resource. Using it does not lead to its depletion.

These properties underscore the essential role of the virtual in the creative process. For Lévy, the passage from the virtual to the actual is not a predetermined, automatic development but the solution to a problem that is not already contained in its formulation: "[Actualization] is the creation, the invention of a form on the basis of a dynamic configuration of forces and finalities. Actualization involves more than simply assigning reality to a possible or selecting from among a predetermined range of choices. It implies the production of new qualities, a transformation of idea, a true becoming that feeds the virtual in turn" (25). As this idea of feedback suggests, the importance of Lévy's treatment of virtuality resides not merely in its insistence on the dynamic nature of actualization but in its conception of creativity as a two-way process involving both a phase of actualization and a phase of virtualization. The complementarity of the two processes is symbolized in Lévy's text by the recurrent image of the Möbius strip, an image that stands in stark contrast to Baudrillard's vision of a fatal attraction toward the virtual.

While actualization is the invention of a concrete solution to answer a need, virtualization is a return from the solution to the original problem. This movement can take two forms. Given a certain solution, the mind can reexamine the problem it was meant to resolve, in order to produce a better solution. Cars, for instance, are a more efficient way to solve the problem of transportation than horse-drawn carriages, but they have multiple drawbacks that could be avoided by better designs or entirely different concepts. Virtualization can also be the process of reopening the field of problems that led to a certain solution and finding related problems to which the solution may be applied. A prime example of this process is the evolution of the computer from a number-crunching automaton to a world-projecting and word-processing machine.

The concept of virtualization is an extremely powerful one. It involves any mental operation that leads from the here and now, the singular, the usable once-and-for-all, and the solidly embodied to the timeless, abstract, general, multiple, versatile, repeatable, ubiquitous, immaterial, and morphologically fluid. Skeptics may object that Lévy's concept of virtualization simply renames well-known mental operations such as abstraction and generalization, but partisans will counter that the notion is much richer because it explains the *mechanisms* of these operations. If thought is the production of models of the world—that is, of the virtual as double—it is through the consideration of the virtual as potential that the mind puts together representations that can act upon the world. While a thought confined to the actual would be reduced to a powerless recording of facts, a thought that places the actual in the infinitely richer context of the virtual as potential gains control over the process of becoming through which the world plays out its destiny.

The power of Lévy's concept of virtualization resides precisely in its dual nature of timeless operation responsible for all of human culture, and of trademark of the contemporary Zeitgeist. In our dealing with the virtual, we are doing what mankind has always done, only more powerfully, consciously, and systematically. The stamp of postmodern culture is its tendency to virtualize the nonvirtual and to virtualize the virtual itself. If we live a "virtual condition," as N. Katherine Hayles has suggested (*How We Became Posthuman,* 18), it is not because we are condemned to the fake but because we have learned to live, work, and play with the fluid, the open, the potential. In contrast to Baudrillard, Lévy does not seem alarmed by this exponentiation of the virtual because he sees it as a productive acceleration of the feedback loop between the virtual and the actual rather than as a loss of territory for the real.

Lévy's examples of virtualization include both elementary cultural activities and contemporary developments. Among the former are tool making and the creation of language. Tool making involves the virtual in a variety of ways. The concrete, manufactured object extends our physical faculties, thus creating a virtual body. It is reusable, thus transcending the here and now of actual existence. Other virtual dimensions of tools are inherent to the design itself: it exists outside space and time; it produces many physically different yet functionally similar objects; it is born of an understanding of the recurrence of a problem (if I need to drive *this* nail here and now, I will need to drive nails in other places and at other times); and it is not worn out by the process of its actualization.

Language originates in a similar need to transcend the particular. The creation of a system of reusable linguistic types (or *langue*) out of an individual or communal experience of the world is a virtualizing process of generalization and conceptualization. In contrast to a proper name, a noun like *cat* can designate not only the same object in different contexts but also different objects in different contexts with different properties: my cat, your cat, the bobcats in the mountains, and the large cats of Africa. It is this recyclable character of linguistic symbols that enables speakers to embrace vast expanses of experience with a finite vocabulary.[3] Whereas the creation of language is the result of the process of virtualization, its use in an act of *parole* is an actualization that turns the types into concrete tokens of slightly variable phonic or graphic substance and binds utterances to particular referents. Even in its manifestation as *parole,* however, language exercises a virtualizing power. Life is lived in real time, as a succession of presents, but through its ability to refer to physically absent objects, language puts consciousness in touch with the past and the future, metamorphoses time into a continuous spread that can be traveled in all directions, and transports the imagination to distant locations.

As examples of more specifically contemporary forms of virtualization, Lévy mentions the transformations currently undergone by the economy and by the human body. In the so-called information age, the most desirable good is no longer solid manufactured objects but knowledge itself, an eminently virtual resource since it is not depleted by use and since its value resides in its potential for creating wealth. On the negative side, the virtualization of the economy has encouraged the pyramid schemes that currently plague the industries of sales and investments. As for the body, it is virtualized by any practice and technology that aims at expanding its sensorium, altering its appearance, or pushing back its biological limits. In a fake-theory of the virtual, the virtualization of the human body is represented by the replacement of body parts with prostheses; it finds its purest manifestation in the implant of artificial organs and cosmetic surgery. In a potential-theory, the virtualization of the body is epitomized by performance- and perception-enhancing devices, such as the running sneaker and the telescope. The inspiration for these practices is the fundamentally virtualizing question, "To what new problems can I apply this available resource, the body I was born with?" as well as the actualizing one, "How should I refashion this body to make it serve these new functions?"

The development of simulation technologies such as VR illustrates yet another tendency of contemporary culture—the virtualization to a second degree of the already virtual. Consider computers. They are virtual objects by virtue of being an idea and a design out of which particular machines can be manufactured. These machines are virtual, as we have seen, in the sense that they can run different software programs that enable them to emulate (and improve on) a number of different other machines. Among their applications are simulative programs whose purpose is to test formal models of objects or processes by exploring the range of situations that can develop out of a given state of affairs. The knowledge gained by trying out the potential enables the user to manage the possible and to control the development of the real. If all tools are virtual entities, computer simulations are doubly or perhaps even triply virtual, since they run on virtual machines and since they incorporate the virtual into their mode of action.

The Text as Double and as Fake

As they are implicated in thought, the two faces of the virtual are also implicated in texts, the inscription and communicable manifestation of the thinking process.

Descriptions of the text, especially of the artistic text, as image functioning as a double of the real go back at least as far as Aristotle.

> Imitation comes naturally to human beings from childhood (and in this they differ from other animals, i.e. in having a strong propensity for imitation and in learning their earliest lessons through imitation); so does the universal pleasure in imitation. What happens in practice is evidence of this: we take delight in viewing the most accurate possible images of objects which themselves cause distress when we see them (e.g. the shapes of the lower species of animals, and corpses). The reason for this is that understanding is extremely pleasant, not just for philosophers but for others too in the same way, despite their limited capacity for it. This is the reason why people take delight in seeing images; what happens is that as they view them they come to understand and work out what each thing is. (*Poetics* 3.1, 6)

We can read this passage as the expression of a classic view of representation, Baudrillard's stage one in the evolution of the image. It is because the work of art provides "understanding" of objects in the world, as it replaces raw sensory experiences with intelligible models of things, that we derive pleasure from the process of artistic duplication. But in stressing human beings' innate propensity for imitation and the "delight" they find in images from early childhood on, the *Poetics* fragment suggests to the modern reader a much less didactic type of gratification: we enjoy images precisely because they are not "the real thing." We enjoy them for the skill with which they are crafted. This pleasure presupposes that the readers or spectators of artistic texts do not fall victim to a mimetic illusion; it is because they know in the back of their minds that the text is a mere double that they appreciate the illusionist effect of the image, the fakeness of the fake.

Baudrillard and Umberto Eco describe this attitude as a typically postmodern attraction for the hyperreal. In his *Travels in Hyper-reality*, for instance, Eco suggests that visitors to Disneyland experience far greater fascination with automata that reproduce pirates or jungle animals than they would with live crocodiles or flesh and blood actors. Both Baudrillard and Eco lament this attraction to the image as a loss of desire for the original. But the Disneyland tourist, beloved scapegoat of cultural critics, deserves credit for the ability to appreciate the art that goes into the production of the fake. Rather than ridiculing the tourist's attitude for its lack of intellectual sophistication, I would suggest that we regard this attitude as the manifestation of a fundamental and timeless dimension of the aesthetic experience.

In the literary domain, the "fake" interpretation of the virtual entertains obvious affinities with the concept of fictionality. The feature of inauthenticity describes not only the irreal character of the reference worlds created by fiction but also, as John Searle has suggested, the logical status of fictional discourse itself. Some literary theorists, most notably Barbara Herrnstein Smith and Mary Louise Pratt, propose to regard fiction as the imitation of a nonfictional genre, such as chronicle, memoir, letter, biography, or autobiography. Without going this far—for many fictional texts do not seem to reproduce any identifiable type of reality-based discourse—we can profitably describe fiction as a virtual account of fact, or, with Searle, as a pretended speech act of assertion, since even though the fictional text evokes imaginary characters and events or attributes imaginary properties to counterparts of real-world individuals, it does so in a language that logically presupposes the actual existence of its reference world.

This idea of the text and, in fact, of the work of art in general as a *virtual something else* has been systematically explored by Susanne K. Langer in *Feeling and Form* (1953). Langer's interpretation of the virtual foregrounds the optical illusion: "The most striking virtual objects in the natural world are optical—perfectly definite visible 'things' that prove to be intangible, such as rainbows and mirages." And also: "An image is, indeed, a purely virtual 'object.' Its importance lies in the fact that we do not use it to guide us to something tangible and practical, but treat it as a complete entity with only *visual* attributes and relations. It has no others; its visible character is its entire being" (48). To extend the optical concept of virtuality to nonvisual forms of art, without resorting to worn-out and medium-insensitive metaphors such as "painting" with words or with sound, Langer detaches the notion of image from any individuated content. Though the work of art is an essentially mimetic text, this mimeticism resides more in the production of an equivalent of one of the fundamental, almost Kantian, a priori categories of human experience than in the reproduction of concrete aspects of life or singular objects. The virtual images of art are not primarily images of bodies, flowers, animals, characters, and events, or the abstract expression of feelings, but what we might call today dynamic simulations of abstract categories of thought, such as space (through visual arts), time (through music), memory (through narrative), and gesture (through dance). And while Langer calls drama virtual history, it could just as well be labeled "virtual action."[4]

The Text as Potentiality

As an analytical concept, the virtual as potential is no less fecund for literary and textual theory than the virtual as fake. Here again we must begin with Aristotle: "The function of the poet is not to say what *has* happened, but to say the kind of things that *would* happen, i.e. what is possible in accordance with probability and necessity" (*Poetics* 5.5, 16). This pronouncement may seem to restrict literature unduly to the representation of events and objects that could occur in the real world, given its physical, logical, and perhaps psychological and economic laws. A narrow interpretation of possibility would leave out not only fairy tales, science fiction, the fantastic, and magical realism but also the absurd, the symbolic, the allegorical, and the dreamlike. All these literary landscapes can be reclaimed by broadening the horizon of "probability and necessity" to the territories covered by a purely imaginative brand of possibility. The task of the poet is not necessarily to explore the alternative worlds that can be put together by playing with the laws of the real but to construct imaginary worlds governed by their own rules. These rules—which may overlap to various degrees with the laws of the real—must be sufficiently consistent to afford the reader a sense of what is and isn't possible in the textual world as well as an appreciation of the imaginative, narrative, and artistic "necessity" of what ends up being actualized.

The virtual as potential also lies at the core of the conception of the text developed by the two leading figures of reader-response criticism, the Polish phenomenologist Roman Ingarden and his German disciple Wolfgang Iser. Ingarden conceives the literary work of art, in its written form, as an incomplete object that must be actualized by the reader into an aesthetic object. This actualization requires of the reader a filling in of gaps and places of indeterminacy that can take a highly personal form, since every reader completes the text on the basis of a different life experience and internalized knowledge. Rather than associating the written or oral signs that make up the text with a specific possible world, it is therefore more appropriate to speak, with David Lewis, of a plurality of textual worlds. In this power to unfold into many worlds resides for Iser the virtuality of the work of art and the condition for the aesthetic experience: "It is the virtuality of the work that gives rise to its dynamic nature, and this in turn is the precondition for the effects that the work calls forth" ("Reading Process," 50).

For Pierre Lévy, the virtual as potential represents not only the mode of being of the literary text but the ontological status of all forms of textuality. "Since its

Mesopotamian origin," writes Lévy, "the text has been a virtual object, abstract, independent of any particular substrate" (*Becoming Virtual,* 47). Paradoxically, this virtual object originates in an actualization of thought. The act of writing taps into, and enriches in return, a reservoir of ideas, memories, metaphors, and linguistic material that contains potentially an infinite number of texts. These resources are textualized through selection, association, and linearization. But if the text is the product of an actualization, it reverts to a virtual mode of existence as soon as the writing is over. From the point of view of the reader, as reader-response theorists have shown, the text is like a musical score waiting to be performed. This potentiality is not just a matter of being open to various interpretations or of forming the object of infinitely many acts of perception; otherwise texts would be no more and no less virtual than works of visual art or things in the world such as rocks and tables. The virtuality of texts and musical scores stems from the complexity of the mediation between what is there, physically, and what is made out of it. Color and form are inherent to pictures and objects, but sound is not inherent to musical scores, nor are thoughts, ideas, and mental representations inherent to the graphic or phonic marks of texts. They must therefore be constructed through an activity far more transformative than interpreting sensory data. In the case of texts, the process of actualization involves not only the process of "filling in the blanks" described by Iser but also simulating in imagination the depicted scenes, characters, and events and spatializing the text by following the threads of various thematic webs, often against the directionality of the linear sequence.

As a generator of potential worlds, interpretations, uses, and experiences, the text is thus always already a virtual object. But the marriage of postmodernism and electronic technology, by producing the freely navigable networks of hypertext, has elevated this built-in virtuality to a higher power. "Thought is actualized in a text and a text in the act of reading (interpretation). Ascending the slope of actualization, the transition to hypertext is a form of virtualization" (Lévy, *Becoming Virtual,* 56). This virtualization of the text matters cognitively only because it involves a virtualization of the act of reading. "Hypertextualization is the opposite of reading in the sense that it produces, from an initial text, a textual reserve and instrument of composition with which the navigator can project a multitude of other texts" (54). In hypertext, a double one-to-many relation creates an additional level of mediation between the text as produced by the author—*engineered* might be a better term—and the text as experienced by the reader. This additional level is the text as displayed on the screen. In a traditional text, we have two levels:

1. The text as collection of signs written by the author
2. The text as constructed (mentally) by the reader

The object of level 1 contains potentially many objects of level 2. In a virtualized text, the levels are three:

1. The text as written or "engineered" by the author
2. The text as presented, displayed, to the reader
3. The text as constructed (mentally) by the reader

In this second scheme, which is also valid for the print implementations of what Eco calls the open work, the textual machinery becomes "a matrix of potential texts, only some of which will be realized through interaction with a user" (Lévy, *Becoming Virtual,* 52). As a virtualization of the already virtual, hypertext is truly a hyper-text, a self-referential reflection of the virtual nature of textuality.

When Lévy speaks of the virtualization of the text, the type of hypertext he has in mind is not so much a "work" constructed by a single mind as the implementation of Vannevar Bush's idea of the Memex: a gigantic and collectively authored database made up of the interconnection and cross-reference of (ideally) all existing texts.[5] It is, properly speaking, the World Wide Web itself. In this database the function of the links is much more clearly navigational than in the standard forms of literary hypertext. The highlighted, link-activating key words capture the topic of the text to be retrieved and enable readers to customize the output to their own needs. In Lévy's words, the screen becomes a new "typereader [*machine à lire*], the place where a reserve of possible information is selectively realized, here and now, for a particular reader. Every act of reading on a computer is a form of publishing, a unique montage" (54). As the user of the electronic reading machine retrieves, cuts, pastes, links, and saves, she regards text as a resource that can be scooped up by the screenful. In a formula that loses a lot in translation, Lévy writes, "Il y a maintenant du texte, comme on dit de l'eau et du sable" (Now there is only text, as one might say of water and sand [62]). If text is a mass substance rather than a discrete object, there is no need to read it in its totality. The reader produced by the electronic reading machine will therefore be more inclined to graze at the surface of texts than to immerse herself in a textual world or to probe the mind of an author. Speaking on behalf of this reader, Lévy writes, "I am no longer interested in what an unknown author thought, but ask that the text make me think, here and now. The virtuality of the text nourishes my actual intelligence" (63).

TWO

Virtual Reality as Dream and as Technology

> Viewing 3-D graphics on a screen is like looking into the ocean from a glass-bottom boat.
>
> We see through a flat window into an animated environment; we experience being on a boat.
>
> Looking into a virtual world using a stereographic screen is like snorkeling. We are at the boundary of a three-dimensional environment, seeing into the depth of the ocean from its edge; we experience being between at the surface of the sea.
>
> Using a stereoscopic HMD [head-mounted display] is like wearing scuba gear and diving into the ocean. Immersing ourselves in the environment, moving among the reefs, listening to the whale song, picking up shells to examine, and conversing with other divers, we invoke our fullest comprehension of the scope of the undersea world. We're There.—*Meredith Bricken*

When virtual reality technology burst into public view in the early 1990s, it was less through a revolutionary computer system than through a grand flourish of rhetoric. The idea of VR sprang fully formed from the brain of its prophets, and the media presented it to the general public as being in a state of perfect implementation. The popular perception of VR was primarily shaped by the declarations of a charismatic developer, as well as musician and visual artist, Jaron Lanier, who coined the term *virtual reality,*[1] and of an imaginative journalist, Howard Rheingold, whose 1991 book *Virtual Reality* took readers on a tour of the underground operations wherein brave new worlds of digital data were rumored to be secretly hatching. Little did many of us realize in the early 1990s that their enthusiastic and precise descriptions of VR applications were largely castles in the air, nor did we realize how much separated "real" VR from the virtual brand. Rheingold's suggestion (345–52) that VR technology might be put in the service of teledildonics, sex in a bodysuit with a computer-simulated partner, did more to put VR on the cultural map than any demonstration of HMD, data gloves, or three-dimensional visual display. But it was an interview with Lanier, published in 1988 in *Whole Earth Review* (reprinted in Zhai, *Get Real*), that provided the most vivid input for the image of the role and potential of VR technology that prevails in the public at large.

Lanier's VR was not a space that we would visit for short periods of time, as is the case with current installations, but a technology that would play a major role in our daily lives and deeply transform the conditions of our material existence. In this sense it would be real rather than virtual. A computer would be installed in our house, the so-called Home Reality Engine; we would turn it on, don a minimal VR outfit—nothing more encumbering than glasses to see and gloves to manipulate—and presto, we would be surrounded by a virtual world in which the material objects that furnish the house would take on whatever appearance we specified. (Projecting virtual appearances onto real objects, rather than creating virtual objects *ex nihilo,* solves the annoying problem of bumping into things.) In this world of our creation, we would take on any identity we wished, but the real body's movements would control the virtual body, and we would interact with the virtual world through physical gestures. The computer would keep track of all our past actions and creations, and since time spent in the system would be a significant part of our lives, these digital archives would become a substitute for memory. We would be able to relive earlier experiences by simply rerunning the software.

VR theorists eagerly endorsed this bold vision. Even scientists adopted Lanier's vision as a goal to shoot for (wasn't he, after all, a computer wizard?)—a situation that often makes it difficult to separate the science-fictional from the scientific and the futurological from the technological in the literature devoted to the idea. The discourse of contemporary culture intertwines at least three points of view on VR, those of dreamers, developers, and philosophers, and in this chapter I attempt to give each its due.

Dreams of VR, and Some Realities

At the first Conference on Cyberspace, held in Austin, Texas, in 1990, imaginations were turned loose and metaphors flowed freely. The term *cyberspace* is now mainly associated with the Internet, but for the participants in the conference it covered a wide range of applications of digital technology that included computer-generated environments—virtual worlds proper—as well as networking. For Marcos Novak, an architect, "cyberspace is poetry inhabited, and to navigate through it is to become a leaf on the wind of a dream" ("Liquid Architecture," 229). Nicole Stenger, an artist and poet using computer technology, declared, "Without exaggeration, cyberspace can be seen as the new bomb, a pacific blaze that will project the imprint of our disembodied selves on the walls of eternity" ("Mind," 51). Michael Heim, well-known VR philosopher, spoke eloquently of the "erotic ontology of cyberspace," widening *erotic* to a Platonic sense ("Erotic," 59). Michael Benedikt, the organizer of the conference and also an architect, described the lure of VR as a timeless attraction to other worlds, insisting on the spiritual and artistic implications of this fascination: "Cyberspace's inherent immateriality and malleability of content provides the most tempting stage for the acting out of mythical realities, realities once 'confined' to drug-enhanced ritual, to theater, painting, books, and such media that are always, in themselves, somewhat less than what they reach, mere gateways. Cyberspace can be seen as an extension, some might say an inevitable extension, of our age-old capacity and need to dwell in fiction" ("Introduction," 6).

This idea of dwelling in fiction evokes a popular theme of recent film and literature: walking into a story and becoming a character. We have seen this theme (known to narratologists as metalepsis) in *Alice in Wonderland* (Alice falls through the hole into narrative scripts that have been running for quite some time), in Woody Allen's story "The Kugelmass Episode" (a professor at an American college steps into the world of *Madame Bovary*), in the philosophical dialogues of Douglas Hofstadter's *Gödel, Escher, Bach* (Achilles and the Tortoise are pushed into the paintings of Escher), and in the 1998 movie *Pleasantville* (two teenagers from the 1990s are transported into their favorite TV show from the 1950s).

Of all the versions of the scenario—or is it a postmodern myth?—none is more familiar to the general public than the Holodeck of the TV series *Star Trek: The Next Generation.* Several authors, including Michael Heim and Janet Murray, have exploited the Holodeck association to give a concrete face to VR. An imagination easily aroused is usually quick to feel let down, and as Lanier observed in late 1998, the Holodeck analogy turned out in the long run to be a double-edged sword for the PR of VR: "As for the waning of virtual reality from public attention, I bear some of the blame for it. I always talked about virtual reality in its ultimate implementation and when that didn't happen, interest declined. Because everyone wanted the Holodeck from *Star Trek,* virtual reality couldn't fulfill its promise so quickly" (quoted in Ditlea, "False Starts"). Precisely because of its utopian character, however, the Holodeck scenario provides a convenient approach to the dreams that were invested in the VR project. In the words of Michael Heim, the Holodeck is "a virtual room that transforms spoken commands into realistic landscapes populated with walking, talking humanoids and detailed artifacts appearing so life-like that they are indistinguishable from reality. The Holodeck is used by the crew of the starship Enterprise to visit faraway times and places such as medieval England and 1920s America. Generally, the Holodeck offers the crew rest and recreation, escape and entertainment on long interstellar voyages" (*Metaphysics*, 122). The scenario of the Holodeck breaks down into the following themes:

1. You enter (*active embodiment*) . . .
2. into a picture (*spatiality of the display*) . . .
3. that represents a complete environment (*sensory diversity*).
4. Though the world of the picture is the product of a digital code, you cannot see the computer (*transparency of the medium*).
5. You can manipulate the objects of the virtual world and interact with its inhabitants just as you would in the real world (*dream of a natural language*).
6. You become a character in the virtual world (*alternative embodiment and role-playing*).
7. Out of your interaction with the virtual world arises a story (*simulation as narrative*).
8. Enacting this plot is a relaxing and pleasurable activity (*VR as a form of art*).

Active Embodiment

Once in a while one hears cultural critics emit the opinion that VR is a disembodying technology, a comment that in our body-obsessed, hedonistic culture amounts to a sweeping dismissal of the project. The chief complaint of these critics (Simon Penny, Anne Balsamo, Arthur Kroker) is that VR replaces the body with a body image, thereby causing a Cartesian mind-body split.[2] This opinion is justifiable if under VR one understands "cyberspace" and the imaginary geography of the Internet, where face-to-face encounters in physical meeting places give way to conversations with strangers lodged in invisible bodies and hiding behind digital avatars. In his period-making novel *Neuromancer*, a work that profoundly shaped the popular conception of VR, William Gibson reinforced the conception of computer technology as hostile to the flesh by insisting on the need to leave the "meat" of the body to reach the Matrix, a global computer network through which the mind enjoys the mystical contemplation of the world translated into a fully intelligible display of digital information.

But as a technology of representation, the Lanier-inspired conception of VR is neither cyberspace, the Internet, nor the product of Gibson's imagination.[3] In the brand of VR that is the concern of this chapter, the participation of the physical body is a primary issue, even when the body is clad in a "smart costume" (i.e., a body image constructed by the system) or when it manipulates a distant puppet through teleoperations. "Our body is our interface," claims William Bricken in a VR manifesto (quoted in Pimentel and Teixeira, *Virtual Reality*, 160). Or as Brenda Laurel argues, VR offers the rare opportunity to "[take] your body with you into worlds of imagination" ("Art and Activism," 14). Compared with walking around town, exploring a virtual world with headset, data gloves, or wired bodysuit may involve a significant loss of corporeal freedom, especially since early systems restricted the reading of the body to head and hand movements, but even in its rudimentary state of development the VR experience allows far more extensive physical action than sitting at a computer terminal and typing on a keyboard.

Spatiality of the Display

For a body to enter into a world, this world must be fully spatial. Lanier, in his pioneering pronouncements about VR, describes the VR experience as follows: "When you put [VR glasses] on you suddenly see a world that surrounds you—you see the virtual world. It's fully three-dimensional and it surrounds you, and as you move your head to look around, the images that you see inside the eyeglasses are shifted in such a way that an illusion is created that while you're moving around the virtual world is standing still" (Zhai, *Get Real*, 176). The idea seems simple enough, but if we take a closer look at this description, we notice that the experience of being inside a computer-generated world involves three distinct components: a sense of being surrounded, a sense of depth, and the possession of a roving point of view. Each of these dimensions improves on, or remediates, a variety of earlier technologies.

The ancestors of the surrounding image are the panorama and the cyclorama, both types of installation that flourished in the nineteenth century. Whereas panoramas were moving pictures that unrolled, like a scroll, between two spindles, so that only a portion of the picture would be displayed at any given time, cycloramas were circular paintings displayed on the walls of an equally circular room, affording the viewer a 360-degree angle of vision (Maloney, "Fly Me," 566). VR combines these two ideas by allowing the body to turn around and inspect various parts of the image, as in cycloramas, and by constantly updating the image, as in mechanical panoramas. The sense of depth created by VR displays is the latest development in a series of mathematical or technological innovations that includes the discovery of perspective in the Renaissance, the stereoscopes of the eighteenth and nineteenth centuries, the Cinerama movies of the 1950s, which conveyed a sense of depth when they were viewed with special glasses, and the large-screen IMAX movies of the present. It takes, however, a movable point of view to acquire a full sense of the depth of an image, because it allows objects to slide within the field of vision and to get bigger or smaller to the viewer as their distance from the eye increases or decreases. (This effect is known as motion parallax.)

Here again VR represents the ultimate achievement in the history of what Jay Bolter and Richard Grusin call "point of view technologies" (*Remediation*, 162). First we had flat depictions that did not project a space beyond their surface and therefore did not assign a point of view to the spectator. We tend to process these representations as "the sign" (or visual icon) of an absent object rather than as its immediate presence in the field of vision. Then we had perspective paintings that extended the pictorial space in front of and behind the canvas. Their two-dimensional projection of a three-dimensional space placed the spectator's body in a fixed location with respect to the depicted object. When a chair in a painting is represented from the right and from above, we will retain this point of view even if we physically move to the left and kneel down to look at the picture from below, though admittedly the effect will not be as dramatic as for a spectator situated at the center of projection.[4] Movies allowed shifts in point of view, as the movements of the camera presented objects from various angles and made them change size for the eye, but the spatial location of the virtual body of the spectator in the movie-world was rigidly determined by the location of the camera. Now imagine that the spectator is able to operate the camera, select the point of view, and maintain a continuous apprehension of the external world. This is exactly what happens when a computer tracks the movements of the user's head and body and updates her vision accordingly. As Frank Biocca and Ben Delaney observe, "With a [headset] the viewer ceases to be a voyeur and comes closer to being an actor in the visual world" ("Immersive Virtual Reality," 68). But it does not take a headset to implement the dimension of roving point of view: any so-called first-person video game on a regular computer screen offers a display that can be navigated with a mouse and that constantly updates itself to reflect the position of the cursor (a substitute for the player's body). In contrast to VR, though, the screen display does not offer three-dimensional stereoscopic effects. VR is the only medium that combines the three properties of 360-degree panoramic picture, three-dimensional display, and a point of view controlled by the user.

Sensory Diversity

Each sense, or faculty, is the target of an art form: literature is for the mind, painting for the eye, music for the ear, cuisine for the taste buds, perfume for the nose. Touch is the hardest to link to art, but even touch can be cajoled by "designed experiences": erotic massage techniques, sculptures meant to be stroked, and carnival rides that induce scary but pleasurable shaking or vibrations. The extraordinary development of media in the twentieth century may be in part responsible, together with the influence of such creators as Richard Wagner and Antonin Artaud (discussed in chapter 9), for the popularity of a conception of total art that insists on the involvement of all the senses in the artistic experience. The closest to this ideal is the opera, with its blend of music, dance, drama, poetry, stage design, costumes, and light effects, but for all its artistic resources, the opera addresses only two of the senses, as do the theater, cinema, and TV.

On a darker note, multisensory experiences have also been cast in the role of anti-art. In *Brave New World,* his dystopic novel of anticipation, Aldous Huxley imagined a society stultified by a kind of movie, the Feely, that offered visual, auditive, olfactory, and tactile stimuli. The spectators could feel every hair of the rug on which the protagonists were making love, every jolt in the crash of a helicopter, and they were so fully absorbed in these sensations that they paid no attention to the silliness of the plot. Despite Huxley's warning that multisensory art would extinguish critical sense and render the imagination obsolete, the idea has retained a powerful hold on the modern mind. In *Finnegans Wake,* as Donald Theall has shown, James Joyce attempted to create a syncretic and synesthetic language that involved the entire sensorium and simulated the effects of all media. The 1950s and 1960s were obsessed with a much more literal but also rather trivial expansion of perceptual dimensions: Cinerama, 3D glasses, movies accompanied by scratch-and-sniff cards, and, to crown it all, Morton Heilig's Sensorama, an arcade-style ride-on machine that simulated a motorcycle ride through New York City in four sensory dimensions: the sight of the Manhattan streets, the roar of the engine and of other traffic, the exhaust fumes of cars and the aroma of pizza cooking in restaurants, and the vibrations of the handlebars (Steuer, "Defining," 43).

Though VR is widely credited with the power to create a richer and more diversified environment than any other medium, its potential contribution to the expansion of the sensory dimensions of an image is really quite limited. It is only through haptic sensations—feeling textures and the resistance of simulated objects—and by enabling the user to grab objects that computers can improve on established technologies of representation. Unfortunately, the simulation of the sense of touch is still in a very primitive stage. The most advanced data gloves provide sensation in only one "finger," though the user is said to adapt quickly to this reconfigured hand because of a phenomenon known in the field as "accommodation to virtual worlds." VR developers have made no serious attempt to include gustatory and olfactory signals, because taste and smell do not lend themselves to computerized simulation; as Biocca and Delaney observe, "Both senses are chemical interfaces with the physical world" ("Immersive Virtual Reality," 96). The addition of rudimentary haptic sensations to the standard repertory of visual and aural data may seem out of proportion with the dream of a complete sensory environment, and in its present stage of development virtual touch does not seem to have much artistic potential, but its significance is perhaps more psychological than purely sensorial. It does not take the simulation of haptic sensations to open virtual worlds, at least virtually, to the sense of touch. The resources of digital imaging make it possible to produce a visual display of such intricate texture and shading that the user feels as if he could reach out and caress the objects. Whether imagined or physically simulated, touch is the sense that conveys the strongest impression of the solidity, otherness, and resistance of an object. As Michael Benedikt has pointed out, the awareness of this resistance is the most fundamental condition of a sense of the real: "What is real always pushes back. Reality always displays a measure of intractability and intransigence. One might even say that 'reality' *is* that which displays intractability and intransigence relative to our will" ("Cyberspace," 160).

Transparency of the Medium

In their book *Remediation*, Bolter and Grusin identify the force that inspires cultures to develop new media as a desire for a total lifelikeness that they call transparency: "Our culture wants both to multiply its media and to erase all traces of mediation: ideally, it wants to erase its media in the very act of multiplying them" (5). But if we could develop a medium that provides a perfect copy of the real, or a perfect illusion of reality, would there still be a need for other media? In its ideal implementation, VR is not merely another step toward transparency that will be "remediated" by future media, but a synthesis of all media that will represent the end of media history. Frank Biocca, Taeyong Kim, and Mark Levy suggest that we may be now witnessing "the early stages of the arrival of the ultimate medium" ("Vision," 13). Lanier, predictably, comes up with the most radical pronouncements: "Virtual Reality starts out as a medium just like television or computers or written languages, but once it gets to be used to a certain degree, it ceases to be a medium and simply becomes another reality that we can inhabit" (Zhai, *Get Real,* 184). Or, "Virtual Reality, by creating a technology that's general enough to be rather like reality was before there was technology, sort of completes a cycle" (187).

In this final chapter of media history, transparency is not an end in itself but the precondition for total immersion in a medium-created world. This explains why Pimentel and Teixeira title the first chapter of their book on VR "The Disappearing Computer." The "virtual reality effect" is the denial of the role of hardware and software (bits, pixels, and binary codes) in the production of what the user experiences as unmediated presence.[5] VR represents in this respect a radical change of direction from the conception of the computer that prevailed when artificial intelligence was the most publicized application of digital technology. In the age of VR, and even more in the age of the World Wide Web, computers are no longer credited with an autonomous mind but serve as pure media—as largely hollow channels for the circulation of information. As Brenda Laurel declares, "Throughout this book [*Computers as Theatre*] I have not argued for the personification of the computer but for its invisibility" (143). Jaron Lanier echoes, "With a VR system you don't see the computer anymore—it's gone. All that's there is you" (Lanier and Biocca, "Insider's View," 166).

This disappearance of the computer represents the culmination of the trend toward more user-friendly interfaces in computer design. Binary coded machine instruction once gave way to the mnemonic letter codes of assembly languages; assembly languages were, in turn, translated into high-level languages with a syntax resembling that of natural languages. Then arbitrary words were supplanted by the motivated signs of icons on the screen. One of the articles of faith of the art of interface design is that the computer is a forbidding object that intimidates the user. Whenever possible, the electronic way of doing things should therefore be explained by metaphors that assimilate the new to the familiar—a strategy reminiscent of the process by which natural languages encode abstract ideas through the transposition of concrete categories. The most famous of these metaphors was the desktop and its assortment of tools represented by icons: pages, files, folders, scissors, glue, erasers, and trash cans. But the icons of the desktop metaphor merely surround the part of the screen where everything happens, and the screen is very much a part of the computer's visible body. For immersion to be complete, visual displays should occupy the entire field of the user's vision rather than forming a world-within-a-world, separated from reality by the frame of the monitor. As Gabriel D. Ofeisch observes, "As long as you can see the screen, you're not in VR. When the screen disappears, and you can see an imaginary scene . . . then you are in VR" (quoted in Pimentel and Teixeira, *Virtual Reality*, 7). In the perfect VR system the disappearance of the computer should be achieved on two levels, the physical and the metaphorical. Physically, the computer will be made invisible to the user by being worn on the surface of the skin as what Lanier, anticipating the smart phones, Apple watches, and Google glasses of today, calls "virtual reality clothing." (Dystopic science fiction warns us of a far more frightening practice, the direct implantation of a computer inside the human body.) Metaphorically, the computer will turn into a space that embraces far more than the desktop and the chat room: this space will be a world for the user to inhabit. "Virtual reality" is not just the ultimate medium, it is the ultimate interface metaphor.

Dream of a Natural Language

The dream of an optimal interface is the dream of a command language naturally fitted to the task at hand. This means that in VR symbolic code must disappear, at least in those areas in which it can be more efficiently replaced by physical actions. According to Jaron Lanier, "There's also the ability of communicating without codes. . . . I'm talking about people using their hands and their mouth, whatever, to create virtual tools to change the content of a virtual world very quickly and in an improvisational way" (Lanier and Biocca, "Insider's View," 160). "So, if you make a house in virtual reality, and there's another person there in the virtual space with you, you have not created a symbol for a house or a code for a house. You've actually made a house. It's that direct creation of reality; that's what I call postsymbolic communication" (161). For Michael Benedikt, this postsymbolic communication signals the beginning of a "postliterate" era in which "language-bound descriptions and semantic games will no longer be required to communicate personal viewpoints, historical events, or technical information. . . . We will become again 'as children' but this time with the power of summoning worlds at will and impressing speedily upon others the particulars of our experience" ("Introduction," 12). Through this language without symbols, people will build a shared reality, and minds will become transparent to each other: "Simply, virtual reality, like writing and mathematics, is a way to represent and communicate what you can imagine with your mind. But it can be more powerful because it doesn't require you to convert your ideas into abstract symbols with restrictive semantic and syntactic rules, and it can be shared by other people" (Pimentel and Teixeira, *Virtual Reality*, 17).

The mystics of ages past—such as Emanuel Swedenborg, the esoteric philosopher of the eighteenth century—had a term for this radically antisemiotic mode of communication. They called it the "language of the angels." It would be easy to dismiss the whole project as vaporous New Age mysticism, but the idea of communication without symbols appears much less angelic if we regard it as a *supplement* to and not as a *replacement* for symbolic expression, and if we interpret *symbol* in the narrow sense proposed by Charles Sanders Peirce: a sign whose meaning is based on a social convention that must be learned by the user. Nonsymbolic does not necessarily mean *post*symbolic, pace Lanier and Benedikt. In many situations symbolic expression is indeed what comes naturally to human agents. We may, for instance, meet people in the virtual world—people real or virtual—and wish to talk to them in French or in English. The Home Reality Engine could also fall into the hands of a hacker who might insist on furnishing his homemade world with very visible virtual computers and on programming them in their native machine language.[6] If, as Susan Brennan observes, "certain actions are more easily done gesturally/spatially (as direct manipulation enthusiasts have noticed)" while "others are more easily done with language" ("Conversation," 403), it would be absurd to exclude symbol-based codes such as language from virtual worlds. Biocca and Delaney give us a more realistic idea of the place of nonsymbolic expression in VR systems by offering concrete examples of its advantages:

> The input devices of highly immersive virtual environments try to conform to the way we interact with the physical world by making use of things such as the movement of our limbs, head, eyes, and other motions in physical space. The difference is best illustrated by an example. Say you want to move a computer graphic representation of a cube. In a nongraphic system you might type: Move cube, Location x = 10, y = 55, z = 42. In virtual reality you simply bend down and pick up the computer graphic cube with your hand and place it on a computer graphic table. The floor, the cube, the table, and the graphic representation of your hand are all data entities in a program, as is the computer's representation of your movement. To you it appears as a naturalistic perceptual event. ("Immersive Virtual Reality," 97)

This ambition to develop natural interfaces within virtual worlds represents a complete turnaround from the philosophy of early structuralism,[7] in which the arbitrary system of signs of what are ironically called "natural" languages was regarded both as the metalanguage into which all other semiotic codes could be translated and as a universal medium whose categories entirely determine how we think and what can be thought. We are now much more open to the idea that thought is not always verbal and that some types of thought are better served by expressive resources that do not involve discrete and arbitrary symbols. Partisans of this view include Jaron Lanier, who named his now defunct company VPL, for Visual Programming Language; Pierre Lévy, who believes that the expressive potential of the computer will be better served by a graphic language that he calls "dynamic ideography" (*L'Idéographie dynamique*) than by alphanumeric symbols; and Brian Rotman, who argues that mathematics should accept diagrams as proofs rather than relying exclusively on reasoning formalized in the traditional symbols of the field. In an advanced VR system, there will be no need for ekphrasis—the verbal description of a visual artwork—because the system will encompass all forms of representation, action, and signification. The multisensory will also be the omnisemiotic.

Alternative Embodiment and Role-Playing

The convenience of virtual environments as a workshop for do-it-yourself bodies serving as manifestations of do-it-yourself identities has been widely extolled, and the legitimacy of these identities endlessly debated by cultural critics. Some critics view the virtual bodies of cyberspace—MOOs (Multi-User Dungeon, Object Oriented), MMORPGs (Massively Multiplayer Online Role-Playing Game), chat rooms, computer games, and VR—as the liberating expression of culturally repressed desires. Others insist that each of us has only one body, located in the real world, and that all this play with virtual bodies and virtual personae does not alter the fact that the only body that really matters is the material body; the self cannot be pried loose from the flesh. Halfway in between are those who maintain that the self is multiple and that digital identities actualize its potential but that all these identities are ultimately supported, held together, or "warranted" (Allucquère Rosanne Stone's expression; see "Will the Real Body Please Stand Up") by the physical body. I am not going to offer here my own solution to these problems, and indeed I have none, because I believe that even in postmodern society, selves are not uniformly diverse, and no theory can speak for the many ways in which we may relate to the virtual bodies and virtual personae that we adopt in cyberspace. It will be sufficient for the present purpose to sketch briefly the importance that role-playing and corporeal involvement have been accorded from the very beginning in the conception of VR.

The possibility of redesigning our bodies and becoming something or somebody else is indeed a central theme in Jaron Lanier's 1988 description of the Home Reality Engine: "The computer that's running the Virtual Reality will use your body's movements to control whatever body you choose to have in Virtual Reality, which might be human or might be something quite different. You might very well be a mountain range or a galaxy or a pebble on the floor. A piano . . ." (Zhai, *Get Real,* 177). What does it mean to become a mountain, pebble, or galaxy, all entities devoid of consciousness? The metamorphoses that Lanier has in mind are not a loss of mental faculties or even a change of personality but primarily a change of point of view and physical abilities. Our virtual bodies may fly or creep on the ground, see everything from high above or put up with the limitations of a terrestrial vision, embrace the whole universe or shrink down to the size of a Lilliputian. Some media theorists who do not have much positive to say about VR praise the technology for teaching the relativity of point of view, a lesson that should inspire empathy since it enables users to experience "what it is like to be something or somebody else" (Bolter and Grusin, *Remediation,* 246).

If we inhabit virtual bodies as a point of view, how will we know what these bodies look like and how will we relate to them? To strengthen the bond between the self and the new body, VR systems may paradoxically dissociate the two, so that users will be able to see an image of their virtual embodiment in a combination of first- and third-person point of view that suggests an out-of-the-body experience. According to Ann Lasko-Harvill, former collaborator with Lanier at VPL, "In virtual reality we can, with disconcerting ease, exchange eyes with another person and see ourselves and the world from their vantage point" ("Identity and Mask," 227). But a mere play with point of view falls short of the free design of identities that Lanier describes above. It is only in MUDs and MOOs, the so-called text-based virtual realities, that users can entirely fabricate their own personae, because all it takes to create a virtual individual in these environments is to post its verbal description on the network. When the reconfiguration of the body depends on technological means, such as headset and data gloves, these means determine the range of possible forms of embodiment available to the user. In the present state of development of VR technology, virtual identities must be selected from a menu of ready-made avatars. The same can be said of those video games in which players create their avatar. Playing a role, in these systems, is not a matter of becoming whomever you want to be but a matter of stepping into what Brenda Laurel has aptly called a "smart costume"—smart because it does not merely alter appearance but implements a change of body dynamics.

Simulation as Narrative

VR is not the static image of something or of nothing that Baudrillard calls a simulacrum but an active system of simulation. For Baudrillard, the essence of simulation is deception: "To simulate is to feign to have what one doesn't have" ("Precession," 3). All of Baudrillard's examples of simulacra are images that deceive by virtue of hiding an absence. The Byzantine icon hides the fact that there is no God. Disneyland hides the fact that the city and country surrounding it are just as unreal as a theme park (though Baudrillard does not bother to explain in what sense Los Angeles and the rest of America are unreal). By the same reasoning VR hides the fact that all reality is virtual. Though these simulacra are fully formed objects, they don't seem to be the product of a creative process, and they don't seem to fulfill a specific purpose. Baudrillard's simulacra are not made, they just are, and they are not *used* to deceive (this would presuppose an agent and an intent). Rather, they *embody* deception as a fundamental cultural and epistemological condition. If they have a function at all, it is to satisfy our need for this condition. Computer simulations differ from this conception of the simulacrum on several essential points. They are processes and not objects; they possess a function, and this function has nothing to do with deception; they are not supposed to re-present what is but to explore what could be; and they are usually produced for the sake of their heuristic value with respect to what they simulate. To simulate, in this case, is to test a model of the world. When the simulated world does not exist, as is the case with the projected uses of Lanier's Home Reality Engine, simulation becomes an autotelic activity, but this does not preclude a heuristic value, since the creation and exploration of imaginary worlds can be an instrument of self-discovery.

The essence of computer simulation, whether in VR or in less sophisticated environments, resides in its dynamic character. Ted Friedman calls simulation a "map-in-time" with a narrative dimension ("Making Sense," 86). A typical simulation consists of a number of agents that are given an environment to live in and some rules to follow. The sum of these elements constitutes a narrative world, complete with characters, setting, and principles of action. Because of its power to model the interaction of many forces and to follow the evolution of a world over a lengthy period of time, computerized simulation is an invaluable tool for the study of complex systems, such as those that form the concern of chaos theory. The simplest simulations comprise only one type of agent; for instance, in the late 1990s, the computer museum in Boston exhibited a simulation of how termites build heaps of wood by picking up scattered pieces. The system starts by distributing termites and pieces of wood randomly on the screen. The termites are given three rules of behavior.

1. Move randomly.
2. If you bump into a piece of wood, pick it up.
3. If you bump into a piece of wood while carrying another one, drop this piece nearby.

After a number of repetitions of this pattern, the wood starts gathering into distinct piles, but since the termites keep gnawing at the edges, the contours of the piles are never definitely outlined, and the piles themselves never solidify into perfect, self-contained shapes. Throw into the system several agents with competing goals—for instance, one species of fish that want to eat other fish and another species that want to swim peacefully in schools—and the narrativity of the system takes on a dramatic shape. If the rules are written in such a way that a goal can be fully attained, the system may even reach a state of equilibrium, the simulative equivalent of narrative closure.

When the system revolves around human input, as is the case in VR and computer games, the simulation becomes the life story of the user or, rather, the story of one of the user's virtual lives in the pursuit of a more or less specific goal. Every action taken by the user is an event in the virtual world. The sum of these events may not present proper dramatic form—an Aristotelian rise and fall of tension—but because all events involve the same participant, they automatically satisfy the looser pattern of the epic or serial (episodic) narrative. A smart system, as we see in chapters 8 and 10, may even steer the user's choices toward Aristotelian structures. There are admittedly no courtroom scribes or sports broadcasters who verbalize everything that happens in the virtual world, so the built-in narrativity of VR is strictly a matter of potentiality. The same can be said of the narrativity of life, or even of the theater, and this is why the expression "untold story," so dear to tabloids, is not necessarily an oxymoron. Drama, life, and VR create narrative material with characters, setting, and actions but without narrators. In contrast to narrated narratives, simulation systems do not re-present lives retrospectively, fashioning a plot in which all events are in the book and all the potential narrative material is available to the storyteller, but instead generate events from a prospective point of view, without knowledge of their outcome. The user lives the story as she writes it through her actions in the real time of a continuously moving present. Taken as a whole, however, a VR system is not merely a nonnarrated narrative but a matrix of doubly possible stories: stories that could be lived, and stories that could be told.

Like a "Garden of Forking Paths"—to parody the title of a short story by Borges that has achieved cult status among theorists of interactive literature—the virtual world is open to all the histories that could develop out of a given situation, and every visit to the system actualizes a different narrative path.

VR as a Form of Art

There is no need to dwell at length on the artistic dimension of VR, since it follows from the successful implementation of its other features. VR technology has a number of practical applications, from flight simulators to remote-control surgery and the exploration of the terrain of distant planets, but from the very beginning it has been the potential of the medium as a tool for creative self-expression that has fascinated its advocates. Through its immersive dimension VR inaugurates a new relation between computers and art. Computers have always been interactive, but until now the power to create a sense of immersion was a prerogative of art. Michael Heim has called VR the "Holy Grail" of the artistic quest: "Rather than control or escape or [merely] entertain or communicate, the ultimate promise of VR may be to transform, to redeem our awareness of reality" (*Metaphysics,* 124). In a pure Platonic spirit, the fulfillment of all the senses will stimulate intellectual faculties and offer an experience that blends the aesthetic with the mystical and metaphysical. Lanier's vision is no less exalted, but rather than drawing on philosophical sources, it is inspired by the intellectual current that runs from Romanticism and Symbolism to Dadaism, Surrealism, and the drug culture of the 1960s. The Home Reality Engine is nothing less than a technological support for the Surrealist/Dadaist ideal of an artwork that transforms daily existence into an aesthetic experience, liberates the creative power of the user, and turns poetry into a way of life: "What's exciting are the frontiers of the imagination, the waves of creativity as people make up new things. . . . I want to make tools for VR that are like musical instruments. You could pick them up and gracefully 'play' reality. You might 'blow' a distant mountain range with an imaginary saxophone" (quoted in Zhai, *Get Real,* 49–50).[8]

Presence, Immersion, and Interactivity

The relative importance of immersion and interactivity in a VR system depends on the system's function. In practical applications, immersion is a means to guarantee the authenticity of the environment and the educational value of the actions taken by the user. In a flight simulator, for instance, the usefulness of the system as a test of what a pilot will do with an actual airplane depends on its power to reproduce the complexity and stressful demands of real flight situations. In artistic applications, by contrast, interactivity tends to be subordinated to immersive ideals. It is because they can act upon the virtual world and because this world reacts to their input that users acquire a sense of its presence. This concept of presence is often used in the technical and semitechnical literature on VR to describe the experience that forms the goal of research in the field: "A virtual reality is defined as a real or simulated environment in which a perceiver experiences telepresence" (Steuer, "Defining," 76). Telepresence—or simply presence, in the VR world—relates to physical presence as virtual reality relates to reality: "Telepresence is the extent to which one feels present in the mediated environment, rather than in the immediate physical environment. . . . This [mediated environment] can be either a temporally or spatially distant *real* environment . . . or an animated but nonexistent *virtual world* synthesized by a computer" (ibid.).

The issue of presence involves two conceptually distinct, though practically related, problems: How do we experience what is *there* as being here (telepresence proper), and how do we experience what is made of information as being material? The answer to these questions breaks down into a technological and a psychological, or phenomenological, problem. Jonathan Steuer observes that on the level of hardware, a system's ability to establish presence is a matter of the depth and breadth of information that it can handle (81). Depth is a function of the resolution of the display, while breadth is dependent on the number of senses addressed by this information. One must assume that in order to create presence, a significant amount of information must be devoted to the production of three-dimensional representation. Presence requires a photorealistic display, with detailed effects of texture and shading, but it does not require a real-world content. Another factor of presence involves the mobility of the user's body with respect to the "present" object. In the real world, an object seen through a window may be just as real as an object that we can touch, but we experience it as far less "present" because the sense of presence of an object arises from the possibility of physical contact with it. The object and the body of the perceiver must be part of the same space.

A theory of presence must therefore incorporate a theory of interactivity. Thomas Sheridan ("Musings," 122) acknowledges this dependency when he lists the following three items as the variables that control the experience of presence:

- Extent of sensory information (a category that covers both depth and breadth)
- Control of relation of sensors to environment (e.g., the "ability of the

observer to modify his viewpoint for visual parallax or visual field, or to reposition his head to modify binaural hearing, or ability to perform haptic search")

- Ability to modify physical environment (e.g., "the extent of motor control to actually change objects")

The first of these factors is responsible for the lifelikeness and three-dimensionality of the display, while the other two represent two distinct modes of interactivity, the ability to explore an environment and the ability to change it. At this point we face two terminological choices: label the product of the first factor *immersivity* and the sum of the three *presence*, or label the first factor *realism* and call the total effect either *immersion* or *presence.* I prefer the second choice because the sense of belonging to a world cannot be complete without the possibility of interacting with it. As for the terms *immersion* and *presence,* they capture two different but ultimately inseparable aspects of the total effect: *im*mersion insists on being *inside* a mass substance, *pre*sence on being *in front of* a well-delineated entity. Immersion thus describes the world as a living space and sustaining environment for the embodied subject, while presence confronts the perceiving subject with individual objects. But we could not feel immersed in a world without a sense of the presence of the objects that furnish it, and objects could not be present to us if they weren't part of the same space as our bodies. This approach means that the factors that determine a system's degree of interactivity also contribute to its performance as an immersive system.

Steuer lists the following factors of immersive interaction, without claiming that the list is exhaustive: "*Speed,* which refers to the rate at which input can be assimilated into the mediated environment; *range,* which refers to the number of possibilities for action at any given time; and *mapping,* which refers to the ability of a system to map its controls to changes in the mediated environment in a natural and predictable manner" ("Defining," 86). The first of these items requires little explanation. The speed of a system is what enables it to respond in real time to the user's actions. Faster response means more actions, and more actions mean more changes. The second factor is equally obvious. The choice of actions is like a set of tools; the larger the set, the more malleable the environment. The factor of mapping imposes constraints on the behavior of the system. The user must be able to foresee to some extent the result of his gestures, otherwise they would be pure movements and not intent-driven actions. If the user of a virtual golf system hits a golf ball, he wants it to land on the ground and not to turn into a bird and disappear in the sky.[9] On the other hand, the predictability of moves should be relative, otherwise there would be no challenge in using the system. Even in real life we cannot calculate all the consequences of our actions. Moreover, predictability conflicts with the range requirement. If the user could choose from a repertory of actions as vast as that of real life, the system would be unable to respond intelligently to most forms of input. The coherence of flight-simulation programs stems, for instance, from the fact that they exclude any choice of activity unrelated to flying. Meaningful interactivity requires a compromise between range and mapping and between discovery and predictability. Like a good narrative plot, VR systems should instill an element of surprise in the fulfillment of expectations.

Technical features such as these explain how digital information systems can connect the user to a virtual world, but if we want to understand in its subjective meaning the experience of "being there," we need a phenomenological approach to the question of virtual presence. As a philosophy of the first-person point of view, of the "being-for" of things rather than of being in itself, phenomenology is uniquely suited to analyze the sense of presence to a world that arises from the inscription of the body in the VR system. In what follows I propose to read VR in the light of the insights of Maurice Merleau-Ponty, a precursor of today's theories of the embodied nature of cognition and on the phenomenology of perception.

The Phenomenological Dimension of the VR Experience

The "there" of VR may not be anywhere, objectively speaking, but since we are supposed to relate to virtual worlds as if they were real, the phenomenological investigation of immersion begins with the investigation of the corresponding experience of "belonging to a world" in real environments. This experience forms the central concern of Merleau-Ponty's major work, *The Phenomenology of Perception.* In this book Merleau-Ponty seeks a compromise between an objectivist ontology that attempts to capture the being of things independently of the observer and a subjectivist stance by which our perception creates objects and endows them with properties. Far from denying the mind-independent existence of the world, Merleau-Ponty focuses on the coming together and mutual determination of the world and consciousness. For the perceiving subject, the world is phenomenal; consciousness assumes its existence because it *appears* to the senses. Moreover, since consciousness is intentional, it apprehends itself as directed toward the world; self-consciousness is thus inseparable from consciousness of the world. *Emergence,* a term made popular by recent cognitive science and the theory of complex systems, describes for Merleau-Ponty the apprehension of things in their alterity: "We must discover the origin of the object at the very center of our experience; we must describe the emergence of being and we must understand how, paradoxically, there *is for us an in-itself*" (*Phenomenology*, 71; italics original). *For us* suggests a subjective stance, but what is for us is an *in itself*, a sense of objective existence.

The conception of consciousness as an intentional act directed toward the world is common to all philosophies affiliated with the phenomenological project. What singles out Merleau-Ponty's thought and makes it particularly relevant to the case of VR is the emphasis he places on the embodied nature of consciousness: "The perceiving mind is an incarnated mind. I have tried, first of all, to reestablish the roots of the mind in its body and in its world, going against the doctrines which treat perception as a simple result of the action of things on our body as well as against those which insist on the autonomy of consciousness. These philosophies commonly forget—in favor of a pure exteriority or of a pure interiority—the insertion of the mind in corporeality" (*Primacy*, 3–4). If consciousness is both incarnate and directed toward the world, the body functions as "point-of-view on the world" (*Phenomenology*, 70) and constitutes "our general medium for having a world" (147). It is by imagining ourselves physically reaching out toward things that we acquire a sense of their presence: "We grasp external space through our bodily situation. A 'corporeal or postural schema' gives us at every moment a global, practical, and implicit notion of the relation between our bodies and things, of our hold in them. A system of possible movements, or 'motor projects,' radiates from us to our environment. Our body is not in space like things; it inhabits or haunts space" (*Primacy*, 5).

The difference between "being in space," like things, and "inhabiting" or "haunting space," like the embodied consciousness, is a matter of both mobility and virtuality. Whereas inert objects, entirely contained in their material bodies, are bound to a fixed location, consciousness can occupy multiple points and points of view, either through the actual movements of its corporeal support or by projecting itself into virtual bodies. The ultimate test of the material existence of things is our ability to perceive them under many angles, to manipulate them, and to feel their resistance. When my actual body cannot walk around an object or grab and lift it, it is the knowledge that my virtual body could do so that gives me a sense of the object's shape, volume, and materiality. Whether actual or virtual, objects are thus present to me because my actual or virtual body can interact with them. In the case of an image, for instance, effects of texture and shading invite the viewer to touch the picture in imagination, thereby creating the corporeal relation that tells her, "This is a real, solid, three-dimensional object that belongs to my world."

Perspective creates a similar effect by suggesting that the depicted objects have a hidden side that could be inspected by a mobile body. This sense of presence can only increase when the technology of representation makes it possible for the physical body to walk around or touch the virtual object, as is the case in VR. For the psychologists Pavel Zahorik and Rick Jenison, the presence of objects in VR is a function of their "possible action relationship to the user, or affordance" (a term coined by the psychologist J. J. Gibson). "Successfully supported action in the environment is a necessary and sufficient condition for presence" ("Presence," 86–87). The ideal VR system is conceived here as an ecology in which every object is a tool that extends the user's body and enables her to participate in the ongoing creation of the virtual world.

In this VR ecology, it is not just individual objects that extend the user's body; the same can be said of the virtual world as a whole. In stark contrast to the extensive family of frozen metaphors that describe space as a container, VR turns space into data that literally flow out of the body. The computer creates the virtual world dynamically by tracking the movements of the head of the user and by generating in real time the display that corresponds to his current point of view. The flesh and blood body of the user is bound to the virtual world by a feedback loop that reads the position of the body as binary data and uses this input to produce the sensory display. Writing about the centrality of the actor's body in modern theater, the drama theorist Stanton B. Garner observes that the field of performance is an environmental space "subjectified and intersubjectified by the physical actors who body forth the space they inhabit" (*Bodied Space*, 3). This metaphor becomes almost literal for the user of VR. In the virtual environment, as in certain shamanistic rituals described by Mircea Eliade, the body stands at the center of the world, and the world irradiates from it. The "lag" that separates the user's movements from the updating of the display in today's imperfect VR systems should act as a reminder of the productive implication of the body in the phenomenal world. Through this generation of space in response to the movements of the body, VR technology offers a dramatization of phenomenological doctrine. As Merleau-Ponty writes, "Far from my body's being for me no more than a fragment of space, there would be no space at all for me if I had no body" (*Phenomenology*, 102). Or again: "By considering the body in movement, we can see better how it inhabits space—and, moreover, time—because movement is not limited to submitting passively to space and time, it actively assumes them" (ibid.). This active engagement of the mobile body with space and time produces a succession of points of view through which the spectacle of the world smoothly unfolds to perception.

> Our own body is in the world as the heart is in the organism: it keeps the visible spectacle constantly alive, it breathes life into it and sustains it inwardly, and with it forms a system. When I walk round my flat, the various aspects in which it presents itself to me could not possibly appear as views of one and the same thing if I did not know that each of them represents the flat seen from one spot or another, and if I were unaware of my own movements, and of my body as retaining its identity through the stages of those movements. (203)

There could be no better evocation of the emergent quality of the VR experience of space than this description of an architectural walk-through. It is no coincidence that one of the major applications of VR, both in its full-body implementation and in its downsized, mouse-operated screen versions, has been the simulation of tours through man-made or natural landscapes such as cities, buildings, campuses, gardens, or imaginary geographies. As David Herman observes in *Story Logic*, the tour offers a dynamic experience of space that contrasts with the static representation of the map. Whereas the map captures a disembodied "god's-eye view" that embraces the entire territory at once, the tour temporalizes the experience of space by revealing it one visual frame at a time. Whereas the map is an abstract model of space, the walk-through is a lived experience. Whereas the map has no direction, the tour traces an oriented path through space.

The same contrast operates between the landscaping philosophy of the formal French gardens of the seventeenth and eighteenth centuries and the nature-imitating (and -improving) design of the English gardens of the Romantic age. With its symmetrical patterns of alleys bordered by meticulously sculptured bushes, the French garden must be seen from an elevated point and contemplated in its totality (both features symbolic of the king's political power); whereas, with its meandering walkways, diverse features (temples, ponds, grottoes), and seemingly random grouping of trees, the English garden must be walked through, and every turn of the path reveals a different landscape. As a static spectacle offered to an omniscient gaze and meant to be apprehended from a fixed perspective, the French garden is the horticultural equivalent of a framed painting; as an emergent landscape choreographed for a wandering eye and a moving body, the English garden is a metaphor for the space management and representation that we find in VR.

This VR relation to space is totally different from what we experience in the "cyberspace" of the Internet. Cyberspace projects not a continuous territory but a relatively loose net made of links and nodes, of routes and destinations, with nothing in between. The destinations, or sites, may be centers of interest, but the connecting routes are not. Travel from site to site is not a voyage through a developing landscape but an instantaneous jump that negates the body, since material bodies can move through space only by traversing it one point at a time. The standard metaphor for cyberspace travel, surfing, gives a false impression of continuity. Rather than riding the crest of a swelling wave, the cybernaut is teletransported to more or less random destinations—the faster, the better—by clicking on hyperlinks. In the nonspace of cyberspace, travel time is wasted time, since there is nothing to see between the nodes. In the simulated space of VR, on the contrary, moving around the virtual world is a self-rewarding activity. If surfing were performed in a fully implemented VR system, we would feel the contour of the wave, rise with it above the water plane, and tumble down at the end of the ride. It would not matter where we ended; the pleasure would be the ride itself, the experience of being carried away by a smooth but mighty force. Even in a system that falls short of full-body immersion, the user can find delight in the sensation of bodily movement that results from the changing perspective on the environment, the growing and shrinking of objects, the pursuit of the horizon. To label VR "cyberspace," or cyberspace "virtual reality," is to confuse kinetics with mere transportation and the making-present of space to the body with its disappearance.

II: The Poetics of Immersion

THREE

The Text as World: Theories of Immersion

> One's memory is apparently made up of millions of [sets of images], which work together on the Identikit principle. The most gifted writers are those who manipulate the memory sets of the reader in such a rich fashion that they create within the mind of the reader an entire world that resonates with the reader's own real emotions. The events are merely taking place on the page, in print, but the emotions are real. Hence the unique feeling when one is "absorbed" in a certain book, "lost" in it.—*Tom Wolfe*

Immersion, in VR, is a technologically induced phenomenon, the experience of being surrounded by data. Immersion in a book, by contrast, is a purely mental phenomenon, the product of an act of imagination. Yet when VR theorists attempt to describe the phenomenon of immersion in a virtual world, the metaphor that imposes itself with the greatest insistence is the reading experience.

> As [users] enter the virtual world, their depth of engagement gradually meanders away from here until they cross the threshold of involvement. Now they are absorbed in the virtual world, similar to being in an engrossing book.
>
> The question isn't whether the created world is as real as the physical world, but whether the created world is real enough for you to suspend your disbelief for a period of time. This is the same mental shift that happens when you get wrapped up in a good novel or become absorbed in playing a computer game. (Pimentel and Teixeira, *Virtual Reality*, 15)

Literary authors have not awaited the development of VR technology to offer their own versions and dramatizations of the phenomenon. Charlotte Brontë conceives immersion as the projection of the reader's body into the textual world: "You shall see them, reader. Step into this neat garden-house on the skirts of Whinbury, walk forward in the little parlour—they are there at dinner. . . . You and I will join the party, see what is to be seen, and hear what is to be heard" (*Shirley,* 9). Joseph Conrad's artistic goal prefigures the emphasis of VR developers on a rich and diversified sensory involvement: "My task which I am trying to achieve is, by the power of the written word, to make you hear, to make you feel—it is, before all, to make you see" (Preface to *Nigger of the Narcissus,* xxvi). For Italo Calvino, the transition from ordinary to textual reality is a solemn event, and it must be marked with proper ceremony. The instructions to the reader that open *If on a Winter's Night a Traveler* suggest the rites of passage through which various cultures mark the crossing of boundaries between the profane and the sacred or between the major stages of life. Opening a book is embarking on a voyage from which one will not return for a very long time: "You are about to begin reading Italo Calvino's new novel, *If on a winter's night a traveler.* Relax. Concentrate. Dispel every other thought. Let the world around you fade. . . . Find the most comfortable position: seated, stretched out, or lying flat. . . . Adjust the light so you won't strain your eyes. Do it now, because once you're absorbed in reading there will be no budging you" (3–4).

Immersion and the "World" Metaphor

The notion of reading as immersive experience is based on a premise so frequently invoked in literary criticism that we tend to forget its metaphorical nature. For immersion to take place, the text must offer an expanse to be immersed within, and this expanse, in a blatantly mixed metaphor, is not an ocean but a textual world. The recent emergence of other analogies for the literary text, such as the text as game (see chap. 5), as network (Landow, *Hypertext;* Bolter, *Writing Space*), or as machinic assemblage (Deleuze and Guattari, *A Thousand Plateaus*), should remind us that "the text as world" is only one possible conceptualization among many others, not a necessary, objective, and literal dimension of literary language, but this relativization should be the occasion for a critical assessment of implications that have too long been taken for granted.

What makes the semantic domain of a text into a world? All texts have a semantic domain, except perhaps those that consist exclusively of meaningless sounds or graphemes, but not all of them construct a world. A semantic domain is the nonenumerable, fuzzy-bordered, occasionally chaotic set of meanings that is projected by (or read into) any given sequence of signs. In a textual world these meanings form a cosmos. "How does a world exist as a world?" asks Michael Heim. "A world is not a collection of fragments, nor even an amalgam of pieces. It is a felt totality or whole." It is "not a collection of things but an active usage that things together, that links them. . . . World makes a web-like totality. . . . World is a total environment or surround space" (*Virtual Realism*, 90–91). For Heim, moreover, worlds are existentially centered around a base we call home. "Home is the node from which we link to other places and other things. . . . Home is the point of action and node of linkage that becomes a thread weaving the multitude of things into a world" (92). Let me sum up the concept of world as having four features: a connected set of objects and individuals, a habitable environment, a reasonably intelligible totality for external observers, and a field of activity for its members.

For the purpose of immersive poetics, a crucial implication of the concept of textual world concerns the function of language. In the metaphor of the text as world, the text is apprehended as a window on something that exists outside language and extends in time and space well beyond the window frame. To speak of a textual world means to draw a distinction between a realm of language—made of names, definite descriptions, sentences, and propositions—and an extralinguistic realm of characters, objects, facts, and states of affairs serving as referents to the linguistic expressions. The idea of textual world presupposes that the reader constructs in imagination a set of language-independent objects, using as a guide the textual declarations but building this always incomplete image into a more vivid representation through the import of information provided by internalized cognitive models, inferential mechanisms, real-life experience, and cultural knowledge, including knowledge derived from other texts. The function of language in this activity is to pick objects in the textual world, to link them with properties, to animate characters and setting—in short, to conjure their presence to the imagination. The world metaphor thus entails a referential or "vertical" conception of meaning that stands in stark contrast to the Saussurian and poststructuralist view of signification as the product of a network of horizontal relations between the terms of a language system. In this vertical conception, language is meant to be traversed toward its referents. Sven Birkerts describes this attitude as follows: "When we are reading a novel we don't, obviously, recall the preceding sentences and paragraphs. In fact we generally don't remember the language at all, unless it's dialogue. For reading is a conversion, a turning of codes into contents" (*Gutenberg Elegies*, 97).

The concrete character of the objects that populate textual worlds limits the applicability of the concept to a category of texts that Félix Martínez-Bonati calls mimetic texts. This term refers to texts devoted to the representation of states of affairs involving individual existents situated in time and space, as opposed to those texts that deal exclusively with universals, abstract ideas, and atemporal categories. We can roughly equate mimetic texts with narrative texts, though their evocation of particular existents does not necessarily fulfill the conditions of closure and coherence that we associate with the notion of plot. Since the class of mimetic texts includes fiction and nonfiction, the notion of textual world does not distinguish the worlds that actually exist outside the text from those that are created by it. Both fictional and nonfictional mimetic texts invite the reader to imagine a world and to imagine it as a physical, autonomous reality furnished with palpable objects and populated by flesh and blood individuals. (How could a world be imagined otherwise?) The difference between fiction and nonfiction is not a matter of displaying the image of a world versus displaying this world itself, since both project a world image, but a matter of the function ascribed to the image. In one case, contemplating the textual world is an end in itself, while in the other, the textual world must be evaluated in terms of its accuracy with respect to an external reference world known to the reader through other channels of information.

The idea of textual world provides the foundation of a poetics of immersion, but we need more materials to build up the project. As we saw in the introduction, poststructuralist literary theory is hostile to the phenomenon because it conflicts with its concept of language (more about this in chap. 6). Reader-response criticism, which should be more open to immersion than any other recent critical school, does not clearly put its finger on the experience, though it often comes tantalizingly close.[1] The building blocks of the project therefore have to be found in the quarries of other fields: cognitive psychology (the metaphors of transportation and being "lost in a book"), analytical philosophy (possible worlds), phenomenology (make-believe), and psychology again (mental simulation).

Transportation and Being "Lost in a Book"

The frozen metaphors of language dramatize the reading experience as an adventure worthy of the most thrilling novel: the reader plunges under the sea (immersion), reaches a foreign land (transportation), is taken prisoner (being caught up in a story, being a *captive* audience), and loses contact with all other realities (being lost in a book). The work of the psychologists Richard Gerrig and Victor Nell follows the thread of these classic metaphors to explore what takes place in the mind of the entranced reader. In his book *Experiencing Narrative Worlds* (10–11),

Gerrig develops the metaphor of transportation into a narrative script that could be regarded as a "folk theory" of immersion.

1. *Someone ("the traveler") is transported . . .* For Gerrig, this statement means not only that the reader is taken into a foreign world but also that the text determines his role in this world, thereby shaping his textual identity.
2. *by some means of transportation . . .* If there are any doubts as to the identity of the vehicle, they should be quickly dispelled by these lines from Emily Dickinson: "There is no Frigate like a Book / To take us Lands away" (quoted in Gerrig, 12, and as epigraph to the whole book).
3. *as a result of performing certain actions.* This point corrects the passivity implicit in the metaphor of transportation and introduces another major metaphor developed in Gerrig's book—reading as performance. The goal of the journey is not a preexisting territory that awaits the traveler on the other side of the ocean but a land that emerges in the course of the trip as the reader executes the textual directions into a "reality model" (Gerrig's term for the mental representation of a textual world). The reader's enjoyment thus depends on his own performance.
4. *The traveler goes some distance from his or her world of origin, . . .* When visiting a textual world, the reader must "do as the Romans do"; he must adapt to the laws of this world, which differ to various degrees from the laws of his native reality. Readers may import knowledge from life experience into the textual world, but the text has the last word in specifying the rules that guide the construction of a valid reality model.
5. *which makes some aspects of the world of origin inaccessible.* This idea can be interpreted in many ways: (a) When the idiosyncratic laws of the textual world take over, we can no longer draw inferences from the real-world principles that were overruled. (b) As is the case with any intense mental activity, a deep absorption in the construction/contemplation of the textual world causes our immediate surroundings and current concerns to disappear from consciousness.
6. *The traveler returns to the world of origin, somewhat changed by the journey.* There is no need to elaborate here on the educational value of reading, even when we read for pure entertainment. In lieu of a theoretical development, let me offer a literary formulation of the same idea: "The

> reader who returns from the open seas of his feelings is no longer the same reader who embarked on that sea only a short while ago" (Pavić, *Dictionary of the Khazars* [female edition], 294).

The best illustrations of this script come from the realm of fiction, but Gerrig's stated purpose is to describe a type of experience that concerns "narrative worlds"—what I would call the worlds of mimetic texts—not just fictional ones. The metaphor of transportation captures how the textual world becomes present to the mind, not how this world relates to the real one, and this sense of presence can be conveyed by narratives told as truth as well as by stories told as fiction. Victor Nell writes that "although fiction is the usual vehicle for ludic reading, it is not its lack of truth—its 'fictivity'—that renders it pleasurable" (*Lost in a Book*, 50). Similarly, it is not the imaginative origin of fictional worlds per se that creates the experience that Gerrig calls transportation. But if a theory of transportation—and, by extension, of immersion—should be kept distinct from a theory of fiction, the two cannot be entirely dissociated, because imaginative participation in the textual world is much more crucial to the aesthetic purpose of fiction than to the practical orientation of most types of nonfiction. While nonfiction sends the reader on a business trip to the textual world, often not caring too much about the quality of the experience—what matters most is what happens after the return home—fiction treats the visit as vacation and mobilizes all the powers of language to strengthen the bond between the visitor and the textual landscape.

Another entangled issue is the relation between immersion and aesthetics. We tend to label a literary work immersive when we take pleasure in it, and we (normally) take pleasure in reading when the text presents aesthetic qualities. But aesthetic value cannot be reduced to immersive power. Poetry is not as immersive as narrative because its relation to a "world" is much more problematic, and among the texts regarded as narrative, some deliberately cultivate a sense of alienation from the textual world or do not allow a world to solidify in the reader's mind. For Gerrig, transportation into a narrative world is not dependent on narrative skills. If I read the word *Texas* in a story, no matter how good or bad the text, I will think about Texas, which means that I will be mentally transported to the place: "Some core of processes is likely to allow readers to experience narrative worlds even when the stories themselves are poorly crafted" (*Experiencing*, 5). In Gerrig's Texas example, however, imaginative transportation to Texas is a consequence of the speech act of reference rather than a consequence of embedding the speech act in a narrative context. We must therefore distinguish a minimal form of transportation—thinking of a concrete object located in a time and place other than our present spatio-temporal coordinates—from a strong form of the experience, by which "thinking of" means imagining not only an object but the world that surrounds it, and imagining ourselves contained in this world, in the presence of this object. The minimal form of transportation is built into language and the cognitive mechanisms of the mind; we cannot avoid it. But the richer forms depend on the resonance in the reader's mind of the aesthetic features of the text: plot, narrative presentation, images, and style.

For Victor Nell, the experience of immersion—or rather, as he calls it, of reading entrancement—is a major source of pleasure but not necessarily a trademark of "high" literary value. *Lost in a Book,* his investigation of the "psychology of reading for pleasure," takes its title from a family of metaphors that present equivalents in many languages: "For example, in Dutch the phrase is 'om in een boek op te gaan'; in German, 'in einem Buch versunken zu sein'; and in French, 'être pris par un livre'"(50). The passivity of these metaphors suggests a smooth passage from physical reality to the textual world. It is indeed in terms of easiness that Toni Morrison describes the experience of a young girl who listens for the umpteenth time to the wondrous story of her birth: "Easily she stepped into the told story that lay before her eyes" (*Beloved,* 29). For a reader to be caught up in a story, the textual world must be accessible through effortless concentration: "In terms of attention theory . . . the ludic reader's absorption may seem as an extreme case of subjectively effortless arousal which owes its *effortlessness* to the automatized nature of the skilled reader's decoding activity" (Nell, *Lost in a Book,* 77–78). Difficult materials hamper immersion because "consciousness is a processing bottleneck, and it is the already comprehended messages . . . that fully engage the receiver's conscious attention" (77). The most immersive texts are therefore often the most familiar ones: "Indeed, the richness of the structure the ludic reader creates in his head may be inversely proportional to the literary power and originality of the reading matter" (ibid.).

But for Nell, the association of immersion with ease of reading is no cause for contempt. Anticipating the objections of elitist literary critics, who tend to judge the greatness of literary works by the standards of the Protestant work ethic—"no pain, no gain"—Nell insists on the importance of immersive reading for both high and low culture. Sophisticated readers learn to appreciate a wide variety of literary experiences, but they never outgrow the simple pleasure of being lost in a book. This pleasure is limiting only if we take it to be the only type of aesthetic gratification. There is no point in denying that the worlds of the stereotyped texts of popular culture are the most favorable to immersion: the reader can bring in more knowledge and sees more expectations fulfilled than in a text that cultivates a sense of estrangement. But immersion can also be the result of a process that involves an element of struggle and discovery. How many of us, after finally turning the last page of a difficult novel, compulsively return to the first page with the exhilarating thought that deciphering is over and the fun can now begin? In literature as in other domains—ballet, music, theater, and sports—it is through hard work that we reach the stage of effortless performance. The most forbidding textual worlds may thus afford the "easy" pleasures of immersion, once the reader has put in the necessary concentration.

To remain pleasurable, the experience of being lost in a book must be temporary and remain distinct from addiction, its harmful relative. Nell describes the difference between immersion and addiction in terms of eating metaphors: addicted readers are "voracious" consumers of books; they devour the text without taking the time to savor it. The story lives entirely in the present, and when the reading is completed, it leaves no residue in memory: "Addictive behavior . . . predicts an underdeveloped capacity for private fantasy" (212). While the addicted reader blocks out reality, the reader capable of pleasurable immersion maintains a split loyalty to the real and the textual world. The ocean is an environment in which we cannot breathe; to survive immersion, we must take oxygen from the surface, stay in touch with reality. J. R. Hilgard has compared the amphibian state of pleasurable entrancement to "dreaming when you know you are dreaming" (quoted in Nell, *Lost in a Book*). Nell explains, "[Hilgard] writes that the observing and participating egos coexist, so that the subject is able to maintain 'a continued limited awareness . . . that what is perceived as real is in some sense not real.' This disjunction, allowing the reader both to be involved and to maintain a safe distance, is neatly captured by her subject Robert, who comments on a movie screen in which a monster enters a cave, trapping a group of children: 'I'm not one of them but I'm trapped with them, and I can feel the fright they feel'" (212–13).

On the basis of these observations, we can distinguish four degrees of absorption in the act of reading:

1. *Concentration.* The type of attention devoted to difficult, nonimmersive works. In this mode, the textual world—if the text projects any—offers so much resistance that the reader remains highly vulnerable to the distracting stimuli of external reality.
2. *Imaginative involvement.* The "split subject" attitude of the reader who

transports herself into the textual world but remains able to contemplate it with aesthetic or epistemological detachment. In the case of narrative fiction, the split reader is attentive both to the speech act of the narrator in the textual world and to the quality of the performance of the author in the real world. In the case of nonfiction, the reader engages emotionally and imaginatively in the represented situation but retains a critical attitude toward the accuracy of the report and the rhetorical devices through which the author defends his version of the events.

3. *Entrancement.* The nonreflexive reading pleasure of the reader so completely caught up in the textual world that she loses sight of anything external to it, including the aesthetic quality of the author's performance or the truth value of the textual statements. It is in this mode that language truly disappears. As Ockert, one of the subjects interviewed by Nell, describes the experience: "The more interesting it gets, the more you get the feeling you're not reading any more, you're not reading words, you're not reading sentences, it's as if you are completely living inside the situation" (290). Despite the depth of the immersive experience, however, this reader remains aware in the back of her mind that she has nothing to fear, because the textual world is not reality.
4. *Addiction.* This category covers two cases: (1) The attitude of the reader who seeks escape from reality but cannot find a home in the textual world because she traverses it too fast and too compulsively to enjoy the landscape. (2) The loss of the capacity to distinguish textual worlds, especially those of fiction, from the actual world. (I call this the Don Quixote syndrome.)[2]

Possible Worlds

What does it mean, in semantic and logical terms, to be transported into the virtual reality of a textual world? The answer is tied to an ontological model that acknowledges a plurality of possible worlds. The fictional worlds of literature may not be, technically speaking, the possible worlds of logicians, but drawing an analogy between the two allows a much-needed sharpening of the informal critical concept of textual world.[3] Originally developed by a group of philosophers including David Lewis, Saul Kripke, and Jaakko Hintikka to solve problems in formal semantics, such as the truth value of counterfactuals, the meaning of the modal operators of necessity and possibility, and the distinction between intension and extension (or sense and reference),[4] the concept of possible worlds has been used to describe the logic of fictionality by Lewis himself and adapted to poetics or narrative semantics by Umberto Eco, Thomas Pavel, Lubomír Doležel, Doreen Maître, Ruth Ronen, Elena Semino, and myself. The applications of possible-world (henceforth, PW) theory to literary criticism have been as diverse as the interpretations given to the concept by philosophers and literary scholars.[5] Since it would be beyond the scope of this section to try to represent the entire movement, I restrict my presentation of PW theory to an approach that is largely my own, even though it is strongly indebted to the pioneering work of Eco, Pavel, and Doležel.[6]

The basis of PW theory is the set-theoretical idea that reality—the sum total of the imaginable—is a universe composed of a plurality of distinct elements, or worlds, and that it is hierarchically structured by the opposition of one well-designated element, which functions as the center of the system, to all the other members of the set. The central element is commonly interpreted as "the actual world" and the satellites as merely possible worlds. For a world to be possible, it must be linked to the center by a so-called accessibility relation. Impossible worlds cluster at the periphery of the system, conceptually part of it—since the possible is defined by contrast with the impossible—and yet unreachable. The boundary between possible and impossible worlds depends on the particular interpretation given to the notion of accessibility relation. The most common interpretation associates possibility with logical laws; every world that respects the principles of noncontradiction and excluded middle is a possible world. Another criterion of possibility is the validity of the physical laws that obtain in real life. On this account, a world in which people can be turned overnight into giant insects is excluded from the realm of the possible. Yet another conceivable interpretation involves the idea of temporal directionality: the actual world is the realm of historical facts, possible worlds are the branches that history could take in the future, and impossible worlds are the branches that history failed to take in the past.

The distinction of the possible from the impossible is a relatively straightforward matter; all it takes is a particular definition of the criteria of accessibility. A much thornier issue is the distinction of the actual from the nonactual within the realm of the possible. Through its centered architecture, PW theory runs into difficulties with postmodern theory. The idea of a world enjoying special status is easily interpreted as hegemonism, logocentrism, negative valorization of the periphery, and a rigid hierarchical organization based on power relations. Another objection frequently heard against the centered model is that even though we all live in the same physical world and share a large number of opinions about its basic furnishing, there is no absolute consensus as to where to draw the boundary between the realm of actually existing objects and the domain of merely thinkable existence. Some of us believe in angels and not UFOs, some of us in UFOs and not angels, some of us in both, and some of us in neither. Moreover, belief is a matter of degree. I may believe weakly in angels, and the borders of my vision of what exists may be fuzzy. According to this argument, it would take a "naive realism" to postulate a singular actual world; for, if reality is incompletely accessible to the mind or not accessible at all, there will be inevitable discrepancies in its representation. Postmodern ideologues may further object that the idea of a unique center ignores the cultural and historical relativity of perceptions of reality. The current emphasis on the value of diversity seems better represented by philosophies that postulate a variety of "world versions" without establishing any hierarchical relations between them, such as the model described by Nelson Goodman in *Ways of Worldmaking*, than by the necessarily centered structure proposed by modal logic.

These objections to the concept of actual world can be circumvented by adopting what David Lewis has called an "indexical" definition of actuality. The opposition between the actual and the possible can be conceived in two ways: absolutely, in terms of origin, or relatively, in terms of point of view. In the absolute characterization, the actual world is the only one that exists independently of the human mind; merely possible worlds are products of mental activities such as dreaming, wishing, forming hypotheses, imagining, and writing down the products of the imagination in the form of fictions. In the relative characterization—the one Lewis advocates—the actual world is the world from which I speak and in which I am immersed, while the nonactual possible worlds are those at which I look from the outside. These worlds are actual from the point of view of their inhabitants.[7] With an indexical definition, the concept of actual world can easily tolerate historical, cultural, and even personal variations. Without sacrificing the idea of an absolutely existing, mind-independent reality, we can relativize the ontological system by placing at its center individual images of reality, rather than reality itself. Most of us conceive the world system as centered because this idea reflects our intuition that there is a difference between fact and mere possibility. An egalitarian model such as Goodman's cannot account for these all-important semantic concepts. But we all organize our private systems around personal representations of what is actual.

I represent this model as shown in figure 1:

- At the center is a hypothetical real world, existing independently of the mind.

- Superposed upon this world of uncertain boundaries are its representations, as held by various individuals or collectively by various cultures. These spheres are the different personal versions of the "absolute" center. Their boundaries overlap because they reflect the same physical reality, and despite the current emphasis on relativity and differences, there is a vast area of consensus as to what exists and what does not.
- Further away, outlined in thinner lines, are the worlds that each of us holds to be possible but nonactual. They stand at various distances from our personal center, depending on how difficult it would be to enact them or on what type of accessibility relations link them to the center. If we interpret possible worlds as textual worlds, the model predicts that for most readers the world of a realistic novel is closer to reality than the world of a fairy tale because its actualization does not require a modification of physical laws. It also predicts that a modern American reader will see a greater discrepancy between reality and the world of Macbeth than a contemporary of Shakespeare would have, because belief in witches was more prevalent in Renaissance England than in the twentieth-century United States.

The applicability of the model to literary theory is not exhausted with the assimilation of textual worlds to possible worlds. In fact, a straight assimilation would be doubly reductive. First, it would obscure the fact that the distinction actual/possible reappears within the semantic domain projected by the text. In the case of mimetic texts, an essential aspect of reading comprehension consists of distinguishing a domain of autonomous facts—what I call the textual actual world—from the domains created by the mental activity of characters: dreaming, hoping, believing, planning, and so on. Mimetic texts project not a single world but an entire modal system, or universe, centered around its own actual world. Second, if nonactual textual worlds were apprehended as mere statements of possibility, there would be no phenomenological difference between counterfactual statements or expressions of wishes, which embed propositions under predicates of nonfactuality, and fictional statements, which, as Lewis observes, take the form of straight assertions of truth.

The concept of immersion is crucially dependent on this distinction. When I process "Napoleon could have won the battle of Waterloo if Grouchy had arrived before Blücher," I look at this world from the standpoint of a world in which Napoleon loses, but if I read in a novel, "Thanks to Grouchy's ability to move quickly

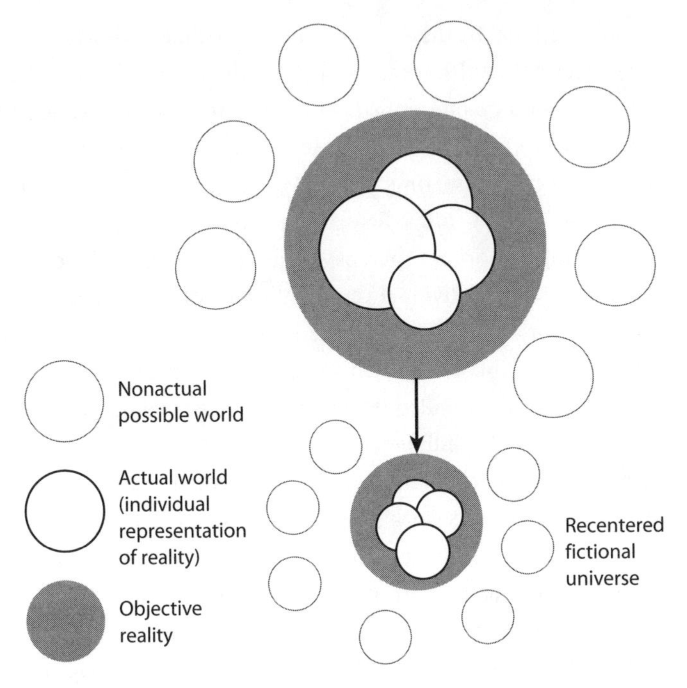

Figure 1. A recenterable possible-worlds model

and bring his army to the battlefield before Blücher, Napoleon crushed his enemies at Waterloo," I transport myself into the textual world and process the sentence as a statement of fact. Both counterfactuals and fictional statements direct our attention toward nonactual possible worlds, but they do so in different modes: counterfactuals function as telescopes, while fiction functions as a space-travel vehicle. In the telescope mode, consciousness remains anchored in its native reality, and possible worlds are contemplated from the outside. In the space-travel mode, consciousness relocates itself to another world and, taking advantage of the indexical definition of actuality, reorganizes the entire universe of being around this virtual reality. I call this move recentering, and I regard it as constitutive of the fictional mode of reading. Insofar as fictional worlds are, objectively speaking, nonactual possible worlds, it takes recentering to experience them as actual—an experience that forms the basic condition for immersive reading.

Recentered universes reproduce the structure of the primary system, except that in the primary system we see only the white circle of our personal actual world, while in recentered systems the reader has access to at least some areas of the patterned circle. In a fictional universe, objective reality corresponds to fictional truths, and fictional truths are established by textual authority. This authority means that fictional truths are unassailable, whereas the facts of the actually actual world can always be questioned. In figure 1 the boundaries of the textual actual world are not clearly defined because individual readers will complete the picture differently and because some texts, especially postmodern ones, leave areas of undecidability or present contradictory versions of facts. (These texts could be represented as having two or more actual worlds, in a blatant violation of the classic modal structure.) The individual representations of reality superposed upon the textual actual world correspond to the personal actual world of characters, while the nonactual possible worlds that surround the center stand for the characters' unfulfilled or partially fulfilled private worlds. Here again, distance from the center stands for degree of fulfillment.

The idea of recentering explains how readers become immersed in a fictional text, but how does the analysis work for texts of nonfiction? It would seem that in this case no recentering is needed because nonfiction describes the real world and the reader is already there, automatically immersed in this "native reality" by some kind of birthright. But where exactly is the reader of nonfiction imaginatively situated: in a text, or in a world? If, as I have suggested, the world-image projected by the text is conceptually different from the world referred to by the text, the reader-persona is located in the reference world, not in its textual image. In fiction, the reference world is inseparable from the image, since it is created by the text, and the contemplation of the image automatically transports the reader into the world it represents. But in nonfiction we can distinguish two moments: (1) one in which the reader constructs the text (i.e., becomes engaged imaginatively in the representation), and (2) one in which the reader evaluates the text (i.e., distances himself from the image, takes it apart, and assesses the accuracy of its individual statements with respect to the reference world). In the first phase, the reader contemplates the textual world from the inside in, and in the second, from the outside in.[8]

The first phase can be more or less elaborate, the reconstructed image more or less vivid and complete, depending on how badly the user needs the textual information for his own practical purposes, but before we can decide what to believe and disbelieve, remember and forget, we must imagine something, and in this act of imagination we are temporarily centered in the textual world. When the textual and the reference world are indistinguishable, as in fiction, the text must be taken as true, since there is no other mode of access to the reference world, and being centered in the textual world implies recentering into the world it represents. When the two worlds are distinct, the image can be true or false, and the reader evaluates it from the point of view of his native reality. The preliminary operation of imaginative centering in this case does not involve ontological recentering. The distinction of a moment of construction from a moment of evaluation avoids two pitfalls frequently encountered in discourse typology: denying any difference in the mode of reading appropriate to fiction and nonfiction, and treating these two modes as incommensurable experiences. It also explains the phenomenon of subjecting one type of text to the mode of reading appropriate for the other. We read fiction as nonfiction when we extract ourselves from its world and, switching reference worlds, assess its viability as a document of real-world events. Conversely, we read nonfiction as fiction when we find the image so compelling that we no longer care about its truth, falsity, or ability to serve practical needs.

Make-Believe

Once we are transported into a textual world, how do we bring it to life? Kendall Walton locates the key to immersion in a behavior that we learn very early in life—earlier, arguably, than we learn to recognize the rigidity of the ontological boundary that separates storyworlds from physical reality. The comparison of fiction to games of make-believe is not a particularly new one; it is implicit to Coleridge's characterization of the attitude of poetry readers as a "willing suspension of disbelief" (*Biographia Literaria,* 169), and it has been invoked by other thinkers, including Susanne K. Langer and John Searle (fiction, for Searle, is "pretended speech acts"). But Walton's project is more ambitious than defining fiction. The stated goal of his book *Mimesis as Make-Believe* is to develop a theory of representation and a phenomenology of art appreciation that make the term *representation* interchangeable with *fiction.* The range of the theory includes not only verbal but visual and mixed media: "In order to understand paintings, plays, films, and novels, we must first look at dolls, hobbyhorses, toy trucks and teddy bears. . . . Indeed, I advocate regarding the activities [that give representational works of art their point] as games of make-believe themselves, and I shall argue that representational works function as props in such games, as dolls and teddy bears serve as props in children's games" (11). The fictionality of all representations is entailed by the definitions that form the axiomatic basis of the project. Here is my own reconstruction of these definitions:

1. A representation is a prop in a game of make-believe.
2. A prop in a game of make-believe is an object—doll, canvas, text—whose function is to prescribe imaginings by generating fictional truths.

3. A fictional truth is a proposition that is "true in a game of make-believe."

Though Walton proposes no formal definition of "game of make-believe"—apparently taking the concept for granted—a set of rules is easily derived from his analysis:

1. Players select an actual object x_1—the prop—and agree to regard it as a virtual object x_2.
2. Players imagine themselves as members of the virtual world in which x_2 is actual. The actions the players perform with the prop count as actions performed with x_2.
3. An action is legal when the behavior it entails is appropriate for the class of objects represented by x_2. A legal action generates a fictional truth.

It is easy to see how these rules apply in the case of children's games. Walton proposes an example in which a group of children decide that stumps are to count as bears. The decision is arbitrary, since any object could be chosen, but once it has been made, the relation between stumps and bears is much stronger than the linguistic relation between the word *bear* and its signified. In the game of make-believe, stumps do not signify absent bears, they are *seen as* present animals. Every time a child sees a stump, she performs an action that counts in make-believe as an encounter with a bear. Players may flee, climb a tree, or shoot the bear but not pet it, put a saddle on its back, or walk it on a leash. The propositions that describe what the stump stands for and what the players' actions count as are the fictional truths. Participating in the game means stepping into a world in which the real-world proposition, "There is a stump," is replaced by the fictional truth, "There is a bear." Every time a player performs a legal move, she makes a contribution to the set of fictional truths that describes the game-world: "I am shooting a bear"; "I am fleeing from it." The pleasure and the point of the game reside in this creative activity.

In visual representation, the stump is the physical image, and the bear is the represented object. The image draws the spectator into its world and confers presence to that which it represents. According to Walton, we behave in front of the painting of a windmill as if we were facing the mill. Inspecting the splotches of color on the canvas counts as inspecting a windmill. The generation of fictional truths is the detection of the visual features of the mill. The legitimacy of moves is determined by the visual properties of the prop, by the nature of the represented object, and by the general rule of the game, which restricts participation to acts of visual perception. Fondling a painting of a nude does not constitute a legitimate response, no matter how erotic the painting's effect may be.

The question "What does the prop stand for?" is slightly more problematic in verbal representation. Assuming that the prop is simply the text, a naive answer could read, "The prop stands for the world it projects." But as Walton observes (*Mimesis*, 219), we may say, "This is a ship," when pointing to the painting *The Shore at Scheveningen* by Willem Van der Velde, but we would never say, "This is a ship," when reading *Moby-Dick*. The difference resides in the fact that while paintings depict iconically, words signify conventionally. The only object that a text can reasonably try to pass as is another text made of the same words but uttered by a different speaker and therefore constituting a different speech act. The basic fictional truth generated by a fictional text is that "it is fictional of the words of a narration that someone [other than the author] speaks or writes them" (356). The prop constituted by the authorial text simply stands for the text of a narrator who tells the story as true fact. The reader performs a game of make-believe that involves three mutually dependent operations: (1) imagining himself as a member of this world, (2) pretending that the propositions asserted by the text are true, and (3) fulfilling the text's prescription to the imagination by constructing a mental image of this world. The range of legitimate actions corresponds to the various world images that can be produced by following the textual directions.

This analysis implies a sharp distinction between texts of fiction and texts of nonfiction. As Walton observes, "It is not the function of biographies, textbooks, and newspaper articles, as such, to serve as props in a game of make-believe." These works are "used to claim truth for certain propositions rather than to make certain propositions fictional" (70). Through a strange asymmetry, however, the distinction "offered for belief" versus "offered for make-believe" is not found in the visual domain. According to Walton, all representational pictures function as props in a game of make-believe, and there is no such thing as nonfictional depiction: "Pictures are fiction by definition" (351). Even pictures primarily used to convey information, such as anatomical illustrations or passport photos, pass as something else and invite the observer to pretend that she is facing that which they represent. All pictures are make-believe because they convey a sense of virtual presence. (Here Walton obviously rejects the idea of a nonillusionist mode of representation, such as we find in pre-Renaissance and postimpressionist art.) The asymmetry between texts and pictures with respect to the dichotomy fiction/nonfiction suggests that fictionality is an essentially verbal category, since without an other to limit and define it, the concept of fiction loses its identity. The asymmetry is partially explained by the fact that pictures do not literally make propositions, but Walton's categorization is above all the consequence of the reinterpretation to which the concept of make-believe is subjected as it crosses the boundary from textual to visual media. In visual communication, make-believe refers to pretended presence: the spectator apprehends the visual features of the depicted object as if she were standing in front of it. In verbal texts, by contrast, make-believe refers to pretended truth for propositions. Walton's use of make-believe thus subsumes and often confuses two distinct phenomena: (1) regarding texts that describe obviously made-up situations as reports of true facts ("willingly suspending disbelief") and (2) engaging in an act of imagination by which depicted objects and their surrounding worlds are made present to the mind. If we disentangle these two aspects of make-believe, the concept is applicable to both the problem of distinguishing fiction from nonfiction (through sense 1) and to the phenomenological description of immersion (through sense 2). While the first sense comes close to being binary, the second is a matter of degree. We can produce sketchy pictures, similar to line drawings, that *signify* objects (i.e., that say "this is a picture of a ship") or rich images, similar to a Vermeer painting, that *show* objects (i.e., that say "this is a ship").[9] Similarly, in the verbal domain, a text can invite readers to consider abstract propositions, or it can invite them to picture scenes vividly in their mind. The immersivity of a text does not depend on its fictionality. As Walton observes, "some histories are written in such a vivid, novelistic style that they almost inevitably induce the reader to imagine what is said, regardless of whether or not he believes it. (Indeed, this may be true of Prescott's *History of the Conquest of Peru.*) If we think of the work as prescribing such a reaction, it serves as a prop in a game of make-believe" (71). Yet an important difference remains between an immersive work of history and an immersive work of fiction. While a fiction fulfills its rhetorical function by producing imaginings, a work of history subordinates imaginings to belief creation. Its ultimate goal is indeed to claim truth, not merely fictional truth, for certain propositions.

Mental Simulation

Immersion is a state, but it takes mental effort to reach it. A favorite term among psychologists to describe this effort is mental simulation. Keith Oatley (*Such Stuff as Dreams*), discussing fiction rather than narrative (but the two largely overlap), takes his lead from computer simulations, such as flight simulators. Stressing the didactic value of fictions (at the cost, in my view, of their entertainment value), he views novels as "simulations of life" that allow readers to explore their emotions in a safe environment, just as users of flight simulators can practice (and learn for real life) dangerous maneuvers without the risk that these actions would entail in an actual situation. In another use of the simulation metaphor, Oatley compares reading to the execution of a computer program that creates simulations by combining two streams of information: one stream provided by the text, which can be regarded as a set of instructions to the imagination, and another provided by the reader, who fills in the gaps on the basis of life experience and cultural knowledge. Since different readers will feed different data into the "program," this metaphor explains both the divergences and relative similarities of interpretations. We can also use the simulation metaphor to describe the computation by the reader of what psychologist Rolf Zwaan calls a "situation model," that is, a mental model of the evolution of the storyworld. In this case the data is provided by the text, backed up by personal experience, and the program is an abstract schema inscribed in the reader's mind that defines the basic conditions of narrativity. The exact form that such a schema should take may be debatable, but if narrative is a universal rather than culture-specific type of meaning, there must be some template in our mind that allows us to produce and understand stories.[10] In the course of reading, we fill out the narrative template, continually updating it with incoming information. The model can be conceived as a state-transition system, in which the states are represented by the properties of the characters and objects that populate a world, and the transitions are represented by the events—actions and happenings—that bring changes in the material world or in the private worlds of characters, such as in their beliefs, wishes, goals, plans, and emotional assessments of the situation.

In another of its psychological uses, the term *mental simulation* is associated with a recent debate concerning the strategies of common-sense reasoning, or "folk psychology." An important aspect of this reasoning is the ability to build a "theory of mind," this is to say, the ability to imagine the thoughts of others with sufficient accuracy to make efficient decisions in interpersonal relations. In contrast to those psychologists who hold that we are able to make judgments about the psychological state of others by activating "a systematically organized body of information about mental states, their origin, interactions and effects" (Heal, "How to Think," 33), a position known as "theory theory," simulationists argue that all we need to do to recreate people's thoughts is to use our existing reasoning abilities with different input—what we take to be the beliefs and values of the foreign mind. According to Stephen Stich and Shaun Nichols, we "take our own decision-making 'off-line,' supply it with 'pretend' inputs that have the same content as the beliefs and desires of the person whose behavior we are concerned with, and let it make a decision on what to do" ("Second Thoughts," 91). Simulation theory can thus be described as a form of counterfactual reasoning by which the subject places himself in another person's mind: "If I were such and such, and if I held beliefs *p* and *q*, I would do *x* and *y*."

Through its implicit shift in point of view, the concept of mental simulation dovetails with the ideas of recentering, transportation, and make-believe, but by locating the reader within the center of consciousness of the characters he tries to understand, it goes further than these concepts in explaining the phenomenon of emotional participation. From a human point of view, one of the most beneficial features of the theory of mental simulation is that it enables us to reason from premises that we normally hold to be false, as well as to gain more tolerance for the thinking processes of people we fundamentally disagree with: "Here the interesting point is that people can think about, and so explore the consequences of and reflect on the interconnections of, states of affairs that they do not believe to obtain" (Heal, "How to Think," 34). Fiction, similarly, has been hailed (and also decried) for its ability to foster understanding of and even attachment to people we normally would condemn, despise, ignore, or never meet in the course of our lives. As we project ourselves into these characters, we may be led to envision actions that we would never face or approve of in real life. As George Eliot observed, "Art is the nearest thing to life; it is a mode of amplifying experience and extending our contact with our fellow men beyond the bounds of our personal life" (quoted from Oatley, "Why Fiction May be Twice as True," 115). Whether we obtain our knowledge of fictional characters' minds through "theory theory" or mental simulation—both ways to construct the mind of real people as well—or through the possibility, arguably unique to fiction, to represent the private thoughts of others, the ability to construct theories of fictional minds has been declared by Lisa Zunshine the reason "why we read fiction" (the title of her 2006 book), for we are natural born voyeurs of private thoughts. While this judgment may be hyperbolic (does it apply to fantasy, thrillers, or science fiction?), accessible minds are certainly a source of immersion, not just for the pleasure of contemplating the passing thoughts of characters and sharing their perceptions, as in Virginia Woolf's novels, but because beliefs, desires, plans, goals, and emotional reactions scaffold the logic of narrative action.

Recent advances in cognitive science have given a neurological foundation to the concept of mental simulation. I am speaking of the "mirror neurons" system, the name given to regions of the brain that have been shown to be activated not only when we perform actions but also when we watch these actions being performed. "Centered on the pre-motor cortex and the inferior parietal lobule, the mirror neuron system is the neurobiological mechanism by which we understand the actions of others and match observed actions with motor representations of these actions. . . . As we observe someone performing an action, our audio-visual areas are activated and our motor system is activated as if we were performing the action ourselves" (Marr, Appendix). This mirroring activity, which can be rightfully called simulation, explains the habit of apes to imitate what they see. It also produces embodied reactions in humans, but it usually stops short of sending the necessary signals to fully reproduce the observed actions. Still, the mental simulation hypothesis goes a long way toward explaining the pleasure we take in watching sports events or in playing computer games. Through a process of identification with the performer, I can become Roger Federer rushing to the net and lobbing a perfect shot into the opponent's court or Mario the plumber performing incredible feats of jumping over abysses.

In its original application, the hypothesis of a mirror-neuron system responsible for mental simulation explains the cognitive processing of actions and movements, but the idea has been extended to other faculties, such as the construction of theories of mind (as described above) and the resulting ability to empathize with others, this is to say, to understand their emotions and "feel their pain" (notwithstanding that the pain of others is normally less painful to us than our own). In all these situations, however, the concept of mental simulation explains our reactions to actually observed phenomena. If the concept is going to account for narrative immersion, it must be shown to work not only in the case of live experience or mediated visual representation (such as film, TV, or computer games) but also in the case of verbal representation. Experiments by Nicole Speer and others at Washington University in Saint Louis indeed suggest that our brain reacts to narrated action in the same way it reacts to performing or observing the same action. The authors took a very simple children's story and presented it a single word at a time to the subjects in order to record their brain states (through fMRI technology) at specific moments. The experiments yielded the following results.

1. When the subjects read about the hero interacting with an object (e.g., "pulled a light cord"), brain activity increased in the areas (the frontal lobes) known to control grasping motions.
2. When the subjects read about character movements through the storyworld (e.g., "went through the front door into the kitchen"), brain

activity increased in areas known to be activated when people view pictures of spatial scenes. This suggests a mapping activity.

3. The signals produced by the subjects were different from the signals produced when reading isolated words. Furthermore, certain regions of the brain were activated only when the state of the storyworld underwent multiple changes between sentences. This suggests that modelization requires a more intense effort.

In an article meant to publicize the research of the faculty of Washington University, journalist Gerry Everding describes the meaning of the experiment: "A new brain-imaging study is shedding light on what it means to 'get lost' in a good book suggesting that readers create vivid mental simulations of the sounds, sights, tastes and movements described in a textual narrative while simultaneously activating brain regions used to process similar experiences in real life" ("Readers").

This interpretation presents a credible view of immersion, but the claims that readers simulate all the senses (sound, sight, taste, and movement) is rather inflated because the experiment only discusses responses to passages involving movement. It seems relatively easy to create mental images of auditory, visual, and kinetic data (especially the latter), but the simulation of olfactory and gustatory sensations is much more problematic. As G. Gabrielle Starr has suggested, it could very well be the case that rather than simulating (imaginatively perceiving) them, we process references to taste and smell semantically, that is, on the basis of meaning and associations ("Multisensory Imagery," 285).

The concept of "simulation" is also problematic in the case of auditory and visual data: Why not just call their mental representation "imagining"? Gregory Currie claims that mental simulation is just another name for an act of imagination ("Imagination," 161). We can however preserve the distinction between imagining and simulating by conceiving the former as the representation of static objects and the latter as the mental images of dynamic objects, as does Ted Friedman (see chapter 2), when he describes simulation as "map in time" ("Making Sense of Software"). In contrast to static representations, the mental picturing of moving bodies involves an embodied simulation, through which the reader identifies with the agent rather than contemplating the moving body from a distant perspective. This may explain why it is much easier to build a mental image of the gestures and actions of characters then to visualize their appearance. I can, for instance, easily picture Emma Bovary sneaking out of her house to join her lover Rodolphe, but I do not have a mental image of her face.

The research of Anežka Kuzmičová, though intuitively rather than experimentally based, confirms the importance of an embodied simulation of movement for narrative immersion. She distinguishes description-imagery, leading to representation, and enactment imagery, leading to what I call simulation. About the latter she writes, "Contrary to common assumption, the reader's body participates in imagining the world(s) of the story to such a degree that bodily movement is frequently emulated from an enactive first-person perspective rather than visualized from the perspective of a passive beholder. As a consequence, references to bodily movement have a unique capacity to make the reader vicariously perceive the world(s) of the story" ("The Words and Worlds," 108). But a text that piles up accounts of character actions and movements does not necessarily produce a greater sense of presence of the storyworld than a text that alternates between description imagery and enactment imagery; on the contrary, an evocation of movement in the middle of a static description may produce in the reader a more intense simulation, leading to an increased sense of presence. Immersivity depends on the artful management of the contrast between representation and simulation rather than on the sheer number of simulation-requiring elements.

Another difference between representation and simulation lies in their direction and heuristic value. In the computer domain, simulations are created to learn about the behavior of a system under different circumstances. In the literary domain, we can distinguish a representational stance that looks backward in time, broadly knowing what is to be represented (though many details will be specified during the act of narration), and a simulative stance that looks forward toward an open future. When we compose a narrative, especially a narrative based on memory, we usually try to represent "how things came to be what they are," and the end is prefigured in the beginning. But when we read a narrative, even one in which the end is presented before the beginning, we adopt the outlook of the characters who are living the plot as their own destiny. Life is lived prospectively and told retrospectively, as Kierkegaard observed, but its narrative replay is once again experienced prospectively. Living a narrative prospectively means placing oneself in a concrete imaginary situation, monitoring its evolution moment by moment, trying to anticipate possible developments, and experiencing the disappearance of possibilities that comes with the passing of time but remaining steadily focused on the hatching of the future. Simulation is the reader's mode of performance of a narrative script.

The term *simulation* may be relatively new, but the idea is an old one. Long before a label was put on the operation, Aristotle recommended its practice to authors of tragedy as a way to ensure the consistency of the plot: "When constructing plots and working them out complete with their linguistic expression, one should as far as possible visualize what is happening. By envisaging things very vividly in this way, as if one were actually present at the events themselves, one can find out what is appropriate, and inconsistencies are least likely to be overlooked" (*Poetics* 8.3, 27). This advice is also valid for writers of narrative fiction. In contrast to narratives of personal experience, novels are often conceived from a prospective stance; the author imagines a situation and tries out many possible developments until a good ending imposes itself. As Currie suggests ("Imagination," 163), the process of world construction is only imperfectly under the conscious control of the creator. While simulating the behavior of characters, the novelist comes to imagine them as autonomous human beings who write the plot for her by taking control of their own destinies. There cannot be a more eloquent tribute to the heuristic value of mental simulation than the feeling voiced by many authors that their characters live a life of their own.[11]

FOUR

Varieties of Immersion: Spatial, Temporal, Emotional

In contemporary culture, moving pictures are the most immersive of all media. Until VR is perfected and becomes widely available, no other form of representation will approximate their ability to combine the spatial extension and fullness of detail of still pictures with the temporality, narrative power, referential mobility (jumping across space and time), and general fluidity of language. This explains why immersion in a book has been compared to "cinema in your head" (Fischlin and Taylor, "Cybertheater," 13). As the reader simulates the story, her mind allegedly becomes the theater of a steady flow of pictures.

How important is the formation of mental images to an immersive reading experience? Do readers construct detailed representations of characters, settings, and actions, something equivalent to a Vermeer painting, or are they satisfied with the schematic outlines created by propositions? The readers who served as subjects in Victor Nell's investigation admit to variable degrees of interest in mental picturing: some describe themselves as "visualizers," some are reading for the plot. "Our imaginings are imprecise and misty," writes William Gass, "and characters in fiction are mostly empty canvas. I have known many who passed through their stories without noses or heads to hold them" (quoted in Nell, *Lost in a Book,* 217). My own experience tells me that novels can occasionally imprint in the mind images of quasi-photographic sharpness, but unlike photographs these images consist of selected features that leave many areas unspecified. I find it relatively easy to imagine settings or moving bodies but virtually impossible to visualize faces. The degree of precision and the nature of the immersed reader's mental representation depend in part on her individual disposition, in part on whether the focus of attention is character, plot, or setting. In this chapter, I take a closer look at the mental operations and textual features responsible for three forms of involvement with narratives: spatial immersion, the response to setting; temporal immersion, the response to story; and emotional immersion, the response to characters.

Spatial Immersion: A Sense of Place and a Model of Space

> It seems plain that the art that speaks most clearly, explicitly, directly, and passionately from its place of origin will remain the longest understood. It is through places that we put up roots. —*Eudora Welty*

> I remember the sensation of reading (Freudians can note this) as one of returning to a warm and safe environment, one that I had complete control over. When I picked up a book it was as much to get back to something as it was to set off to the new.—*Sven Birkerts*

The contrast between space and place, developed by Yi-Fu Tuan, is one of the most celebrated pairs of concepts in human geography. The opposition is one of abstraction versus concrete environment invested with emotional value. Space is infinite, while place is limited by boundaries. Space allows movement—and therefore freedom, adventure, and danger—while place offers a sense of security but also of containment. Space is timeless, while place is shaped by history and memory; space is anonymous, while place is associated with community.

Place and narrative have natural affinities. It is through the stories that are told about them—whether legends, historical facts, or personal memories—that we develop affective bonds with certain places. One could in fact say that locations acquire their status of being places through the stories that single them out from the surrounding space. Conversely, it is quite often because of their ability to evoke places we love that we select the stories we read.

A narrative's ability to immerse readers in place is often the result of a "madeleine effect" that depends more on the coincidental resonance of the text with the reader's personal memories than on generalizable textual properties. Just as the taste and smell of a piece of madeleine dipped into a cup of tea took Marcel Proust's narrator back to the village of his childhood, a single word, a name, or an image is often all the reader needs to be transported into a cherished landscape—or into an initially hated one that grew close to the heart with the passing of time. This phenomenon is documented by the reaction of Gregory Ulmer to this sentence from Michael Joyce's *Twelve Blue*: "Blue isn't anything. Think of lilacs when they are gone." "It so happens," writes Ulmer, "that I never stopped thinking of the lilacs that grew in the backyard of my childhood home, the very scarcity of flowering bushes in Montana making their brief but fragrant appearance all the more impressive. I am hooked" ("Response," para. 2). Such comments are usually judged too impressionistic to be taken seriously by literary theory, but they reveal a dimension of the phenomenology of reading that cannot be ignored.

For all the affinities between storytelling and sense of place, language is not an ideal medium to convey the presence of a place. While a picture transports the spectator almost instantly into a landscape, language can only describe it detail by detail, bringing it slowly into the reader's mind. How can a literary work capture the feel of a place in both its atmosphere and its topography without losing the reader in a descriptive thicket? One solution, proposed by Balzac, is to tie description to a moving body. Balzac's novels open with meticulous evocations of the setting, but the descriptions never jeopardize the reader's sense of orientation because they trace a precise itinerary through the fictional world. When the novel describes a house, such as the boardinghouse of Mme. Vauquer in *Père Goriot* or the decrepit manor in "La Grande Bretèche," the narrator inspects the building in a systematic manner, approaching it from the street, examining the garden and facade, entering through the main door, and walking from room to room, as would a real estate agent or a prospective tenant. The reader ends up with a precise notion of the configuration of the building, all the way down to the floor plan.

To dramatize the description, Balzac often resorts to the device of figuratively pulling the reader into the scene through a second-person address. The four-page depiction of the provincial town of Saumur that opens *Eugénie Grandet* takes readers on a walk up a narrow cobbled street, lets them peek into the backyards, invites them to browse in the stores ("Entrez"), and finally ushers them into the house where the action is to take place. As it weaves its way through the town of Saumur, the descriptive itinerary creates a narrative thread that facilitates the recalling of the images disposed along the way, thus building a "memory palace" comparable in effect to the mnemonic techniques of the sixteenth century: "When you have followed the windings of this impressive street whose every turn awakens memories of the past, and whose atmosphere plunges you irresistibly into a kind of dream, you notice a gloomy recess in the middle of which you may dimly discern the door of Monsieur Grandet's house. Monsieur Grandet's house! You cannot possibly understand what these words convey to the provincial mind unless you have heard the story of Monsieur Grandet's life" (37).

The immersive quality of Balzac's descriptions is measured not by the degree of absorption they arouse in the reader at the time of their reading but by their lingering effect on the rest of the novel. Many people find the beginning of *Eugénie Gran-*

det exasperating rather than immersive. We may indeed hurry impatiently through the description of Saumur, eager for the real action to begin, but the atmosphere that has been fixed in the first few pages will facilitate the process of mental simulation and enrich our mental representation of all the episodes to come.

Balzac's habit of establishing the setting all at once, at the beginning of every novel, reflects his deterministic belief in the importance of the environment for the development of the individual. In Emily Brontë's *Wuthering Heights,* by contrast, the setting is constructed throughout the novel, in delicate and brief strokes, and it seems to emanate from the characters rather than the other way around. Long after readers have forgotten the details of the plot of *Wuthering Heights,* they retain the landscape in their minds; yet the novel hardly ever pauses to give a detailed description of the environment. The sense of place is created dynamically by a narration focalized through the character who is being followed. While the movements of the characters between the two houses of Thrushcross Grange and Wuthering Heights map the geography, their thoughts and perceptions condense the atmosphere:

> [Told as Lockwood, the narrator, enters the court of Heathcliff's house.] Wuthering Heights is the name of Mr. Heathcliff's dwelling. "Wuthering" being a significant provincial adjective, descriptive of the atmospheric tumult to which its station is exposed in stormy weather. Pure, bracing ventilation they must have up here at all times, indeed: one may guess the power of the north wind blowing over the edge, by the excessive slant of the few stunted firs at the end of the house; and by the range of gaunt thorns all stretching their limbs one way, as if craving alms of the sun. (2)
>
> Yesterday afternoon set in misty and cold. I had half a mind to spend it by my study fire, instead of wading through heath and mud to Wuthering Heights. . . . [Lockwood decides to go anyway.] On that black hilltop the earth was hard with a black frost, and the air made me shiver through every limb. (6)
>
> One time I passed the old gate out of my way, on a journey to Gimmerton. It was about the period that my narrative has reached: a bright frosty afternoon; the ground bare, and the road hard and dry. (99)

Through the quasi-instantaneous snapshots of these "narrativized descriptions," as Harold Mosher and others call the technique, the problem of segmentation is minimized, and the experience of the landscape blends with the forward movement of time. The sense of the presence of the environment is out of proportion with the diversity of its features: landscape in *Wuthering Heights* is reduced to a few recurring motifs, such as the wind on the moor, the hard frozen ground in the winter, the soft waves of the grasses in the summer. This economy of detail conveys the vast emptiness of the environment, but it also suggests that textual worlds, like dreamscapes, need only a few mooring points to take hold of the mind, especially when they are already inscribed in the imagination as what Gaston Bachelard calls *rêverie des éléments* (elemental imagination). More than the evocation of a specific English province, landscape in *Wuthering Heights* is a dialogue of earth and wind, an archetypal confrontation of cosmic elements.

A particularly efficient way to create a sense of place without resorting to lengthy descriptions is the use of proper names. From a semantic point of view, proper names contrast with common nouns through their intrinsic lack of sense and the uniqueness of their reference: in a perfect nomenclature, every object in the world would have a different name. The function of names is not to designate the properties of a certain object but to call its existence to the attention of the hearer, to impose it as discourse topic—in short, to conjure a presence to the mind. Through the instantaneous character of the act of reference, the use of a place name teletransports the reader to the corresponding location. For Richard Gerrig, as I noted in chapter 3, the mere mention of the name Texas in a novel lands the reader in Texas or, rather, lands Texas in the mind of the reader. Names may be technically void of sense, but they make up for this emptiness through the richness of their connotations. The name Texas transports the reader not into a barren expanse but into a territory richly landscaped by cultural associations, literary evocations, personal memories, and encyclopedic knowledge. Through this ability to tap into reservoirs of ready-made pictures, place names offer compressed images and descriptive shortcuts that emulate the instantaneous character of immersion in the space of visual media.

From an imaginative and ontological point of view, the place names of fictional worlds fall into several categories. The popularity of regional literature and the predilection of many readers for stories taking place in familiar locations suggest that the most immersive toponyms are the names of real places, either well known or obscure, that we happen to have personally visited, because it is always easier to build mental representations from materials provided by personal experience than by putting together culturally transmitted images—photographs, paintings, movie shots—or by following the instructions of purely textual descriptions.

Next on the scale of immersivity are the names of famous real places we have heard of and dreamed about but never visited. (Readers whose imaginations are more oriented toward culture and history than toward space and landscape will probably invert my rankings.) Western culture elevates locations such as Paris, Venice, Vienna, Provence, New York, or California to mythical status, and these names function for most people as catalysts of desire. Proust has eloquently described the magic of such names:

> I need only, to make [these dreams of the Atlantic and of Italy] reappear, pronounce the names Balbec,[1] Venice, Florence, within whose syllables had gradually accumulated the longing inspired in me by the places for which they stood. . . . But if these names permanently absorbed the image I had formed of these towns, it was only by transforming that image, by subordinating its reappearance in me of their own special laws; and in consequence of this they made it more beautiful, but at the same time more different from anything that the towns of Normandy or Tuscany could in reality be, and, by increasing the arbitrary delights of my imagination, aggravated the disenchantment that was in store for me when I set out upon my travels. (*Remembrance*, 420)

These quasi-mythical sites are often surrounded in fictional worlds by obscure real place names that stand for an entire category of nondescript provincial towns. The Paris of Balzac and Flaubert is irreplaceable, but it wouldn't matter much to most French readers if the Saumur of *Eugénie Grandet* were named Troyes or the Rouen of *Madame Bovary* became Nantes or Bayeux. The place name, in this case, represents a stereotype, and readers construct the setting by activating the cognitive frame with which the text associates the name: "provincial town," "fishing village," "slum," "industrial zone," "vacation resort," and so on. If we can use our idea of French provincial towns to imagine the fictional counterpart of the real Saumur, we can similarly activate our conception of American suburbs to visualize an invented Springfield or Glendale, or, to remain in a purely literary domain, we can draw from a standardized "generic landscape," enriched by personal fantasies of idyllic settings, to picture the Arcadia and the *loci amoeni* of pastoral romance. Well-chosen imaginary place names that conform to the toponymy of a certain region are just as efficient at conveying *couleur locale* as the names of actual locations. Proust's invented Combray, Méséglise, Martinville, or the river Vivonne exude for me the same French *saveur de terroir* (earthy flavor) as Gerard de Nerval's Ermenonville, Châalis, or Loisy—all real names of the province of Valois mentioned in his novella *Sylvie.*

It would be hard to verify experimentally the immersive power of topographical names, but I can illustrate it by a personal anecdote. When I read *Einstein's Dreams* by Alan Lightman, I felt myself literally transported to the city of Bern. I could clearly visualize the bend of the Aar in which the city is built, the many bridges that span the river, the Alps in the background. I could accompany Einstein on his fishing trips, admire the view, feel the current, and I could follow him in his wandering in the old town. The descriptions seemed so accurate, so precise! But this precision was based on a scattering of place names, especially street names, which do not require of the writer to have actually visited the area. When I met the author and told him what a great sense of place the book conveys, he told me that he had never been in Bern and had no intention of going there, so as to preserve in his mind the mythical aura of the city where Einstein conceived his greatest ideas.

Another place-constructing device that shortcuts the linearity of language is what Tom Wolfe calls the detailing of status life and Roland Barthes ascribes to *l'effet de réel* (the reality effect): the mention of concrete details whose sole purpose is to fix an atmosphere and to jog the reader's memory. For the trivial to exercise its signifying function, it must appear randomly chosen and be deprived of symbolic or plot-functional importance. Intrigued by the mention in Flaubert's tale "Un Coeur simple" of the barometer and pyramid of boxes and cartons in Mme. Aubain's room, Barthes asks the question, "Is everything in the narrative meaningful, significant? And if not, if there exist insignificant stretches, what is, so to speak, the ultimate significance of this insignificance?" ("Reality Effect," 12). The ultimate function of such details, according to Barthes, is to tell the reader, "This is the real world." But the device is not merely a convention of realistic fiction. The profusion of details with which J. R. R. Tolkien imagined Middle Earth is a strong factor in the cult that his fantastic world generated. The reader's sense of being there is independent of the verisimilitude of the textual world.

Conveying a mental model of space is a much more difficult task than conjuring the presence of a place because language, as a temporal medium, presents spatial features one after the other. As these features chase each other in memory, it is difficult for the mind to figure out how they relate spatially to each other. In Alain Robbe-Grillet's *In the Labyrinth*, for instance, setting is painstakingly described through a linear accumulation of details, but details flow by the reader's mind without forming a stable geography. The text does not fail to achieve but rather actively *inhibits* spatial immersion:

> I am alone here now, under cover. Outside it is raining, outside you walk through the rain with your head down, shielding your eyes with one hand while you stare ahead nevertheless, a few yards ahead, at a few yards of wet asphalt; outside it is cold; the wind blows between the bare black branches; the wind blows

> through the leaves, rocking whole boughs, rocking them, rocking, their shadows swaying across the white roughcast walls. Outside the sun is shining, there is no tree, no bush to cast a shadow, and you walk under the sun shielding your eyes with one hand while you stare ahead, only a few yards in front of you, at a few yards of dusty asphalt where the wind makes patterns of parallel lines, forks, and spirals. (141)

This passage creates a strong sense of atmosphere, but the incantatory tone of the description, its numerous repetitions, and its paratactic accumulation of details have such a dulling effect that some readers may fail to notice the abrupt switch from rain to sun and from winter to summer. If noticed, however, the transformation should lead to an even greater sense of disorientation. In order to support such a discontinuity, the textual universe cannot be a homogeneous Cartesian space with stable reference points but a self-transforming expanse riddled with invisible black holes through which we are unknowingly sucked into parallel worlds. This conception of space is more hostile to immersion than a mental fog that conceals contradictions, because the imagination presupposes the container of a Cartesian space for the shapes of objects to be representable at all.

One conceivable way to gain a global view of the spatial organization of a storyworld is to simulate mentally the movements of characters. As J. Hillis Miller writes, "A novel is a figurative mapping. The story traces out diachronically the movement of the characters from house to house and from time to time, as the crisscross of their relationships gradually creates an imaginary space. . . . The houses, roads, and walls stand not so much for the individual characters as for the dynamic field of relations among them" (*Topographies*, 19–20). Yet the mapping of the storyworld through the simulation of character movements has its limits, because the visualizations produced by readers focus on individual scenes, and they do not easily coalesce into a global vision of narrative space. But forming such a vision is rarely required for the understanding of plot. When the spatial configuration of the storyworld truly matters, as in crime stories, no verbal description can beat a graphic map. A case in point is the map of Treasure Island in Robert Louis Stevenson's novel. While the plot could be followed from the text alone, the map makes the story much more immersive because it helps the reader understand the importance of the configuration of the island for the characters' strategic decisions.

There are some (rare) moments in literature when space is experienced not merely as a particular geography but in its abstract nature, as that which allows movement and affects perception. In film this effect is reached by mounting the camera on a moving support; in computer games, by altering the display according to the avatar's evolving location in the game world. A famous literary example of this kind of spatial experience is the passage in Proust's novel where the narrator describes how the steeples of Martinville alter their position with respect to each other as the car in which he is riding moves across the countryside. The passage is too long to quote, but here is another passage, from W. G. Sebald's *The Emigrants,* that presents space as the disclosure of landscape from a dynamic perspective. Here the road has replaced the car as the moving object from which the world is seen.

> From Bad Kissingen the road to Steinach goes by way of Grossenbrach, Kleinbrach, and Aschach with its castle and Graf Luxburg's brewery. From there it climbs the steep Aschacher Leite, where Lazarus (Luisa writes) always got down from his calèche so that the horses would not have so hard a job of it. From the top, the road runs down, along the edge of the woods to Höhn, where the fields open out and the hills of the Rhön can be seen in the distance. The Saale meadows spread before you, the Windheim woods nestle in a gentle curve, and there are the tip of the church tower and the old castle—Steinach! Now the road crosses the stream and enters the village, up to the square by the inn, then down to the right to the lower part of the village, which Luisa calls her real home. (194)

Spatio-Temporal Immersion: How to Transport the Reader onto the Scene

From a logical point of view, the narrator and narratorial audience of a story told by the narrator as true fact are located in the textual reference world,[2] but this (re) location does not necessarily land them *on the scene* and *at the time* of the narrative window—to the heart of the storyworld. One of the most variable parameters of narrative art is the imaginative distance between the position of narrator and addressee and the time and place of the narrated events. Spatio-temporal immersion takes place when this distance is reduced to near zero.

Three examples illustrate different degrees of reader proximity to the narrative scene and different strategies to reduce the distance. First, take Boccaccio: "I say that in the city of Pistoia, there was once a very beautiful widow, of whom, as chance would have it, two of our fellow-Florentines, who were living in Pistoia after being banished from Florence, became deeply enamoured" (*Decameron,* ninth day, first story, 682). This passage conforms to what Mary Louise Pratt describes as the standard "natural" (i.e., real-world) storytelling situation, in which a narrator informs an audience of events that took place at a temporal and spatial distance from the present location, the narrator knows the facts, and he displays their report for the entertainment and/or information of the audience.[3] These parameters are confirmed by the framing tale of the *Decameron:* ten young people locked up in a church during an outbreak of the plague, telling each other stories to entertain themselves during their confinement. While the narrator and his audience are located in the same discursive space—in this case, the storytelling event in the church—neither is part of the spatial and temporal window occupied by the narrated events, and neither perceives these events through the senses of the body. This particular passage verifies Seymour Chatman's description of the epistemological foundations of narration: "The narrator can only report events: he does not literally 'see' them at the moment of speaking them. The heterodiegetic narrator never saw the events because he/she/it never occupied the storyworld" (*Coming to Terms,* 144–45).[4]

Chatman proposes this statement as a general model of narration, but the limits of this account are demonstrated by this passage from *Madame Bovary*:

> The bedroom, as [Homais and Dr. Canivet] entered, was mournful and solemn. On the sewing table, now covered with a white napkin, were five or six small wads of cotton in a silver dish, and nearby a large crucifix between two lighted candelabra. Emma lay with her chin sunk in her breast, her eyelids unnaturally wide apart; and her poor hands picked at the sheets in the ghastly and poignant way of the dying, who seem impatient to cover themselves with their shrouds. Pale as a statue, his eyes red as coals, but no longer weeping, Charles stood facing her at the foot of the bed; the priest, on one knee, mumbled under his breath. (367–68)

This episode combines several acts of consciousness: the view offered to the visitors who enter the room; the sensory perception of an invisible observer located on the scene; and the general reflections of an authorial figure about the habits of the dying. These various perspectives blend so smoothly that it almost seems that the events inscribe themselves as they occur in a recording mind. The backgrounding of the act of telling annihilates the imaginative distance between discursive space and storyworld, fusing the consciousness of reader and narrator into the same act of perception. The virtual body whose perspective determines what is perceived belongs at the same time to the narrator and the reader—or to be more precise, to the reader's counterpart in the fictional world—just as, in classical paintings, the eye that contemplates the scene belongs to both painter and spectator. But there is more to immersion than projecting one's virtual body into the fictional world and onto the scene of the events. For the sense of presence to reach its highest intensity, the virtual body must participate actively in the scene. This immersive climax is induced by Emma's gesture: "her poor hands picked at the sheets in the ghastly and poignant way of the dying, who seem impatient to cover themselves with their shroud." According to Anežka Kuzmičová, who describes another Flaubert passage ("The Words and Worlds," 114), when the reader reaches this point she is no longer a passive observer but emulates (or simulates) Emma's gesture from a first-person enactive perspective. Some readers may unconsciously stretch their fingers and try to bring a sheet over their face when reading this passage. It is the contrast between the static propositions of the description and the simulation-inducing gesture (not to mention the dramatic comparison of the sheet with a shroud) that gives the narration of the deathbed scene its immersive power.

In this passage from James Joyce's short story "Eveline," we observe a similar fusion of the virtual body of the narrator and reader with the fictionally real body of a member of the textual world:

> She sat at the window watching the evening invade the avenue. Her head was leaned against the window curtains and in her nostrils was the odour of dusty cretonne. She was tired.
>
> Few people passed. The man out of the last house passed on his way home; she heard his footsteps clacking along the concrete pavement and afterwards crunching on the cinder path before the new red houses. (36)

The reader does not watch a narrator watching Eveline watch the street through the window but, by virtue of the transitivity of the representation of mental processes, directly perceives Eveline's perception. Through identification with the body of Eveline, the reader gains a solid foothold on the scene, as well as a sensory interface to the textual world. The narrative scene becomes as close to the reader as the smell of dusty cretonne to Eveline's nostrils or the texture of the fabric to her cheeks.

The imaginative transportation of the reader's virtual body into the scene of the events is facilitated by a variety of narrative strategies that are often used in contrast with another device. These contrasts include (1) scene versus summary, (2) internal and variable focalization (representing characters as subjects) versus external focalization (looking at characters as objects), (3) dialogue and free indirect discourse bearing the marks of the characters' idiosyncrasies versus stylistically neutral indirect reports of speech, (4) prospective first-person narration representing the textual world from the point of view of the narrator-then (as hero of the tale) versus retrospective representation informed by the knowledge of the narrator-now (as historian of his own life), (5) totally effaced or aggressively visible "hectoring" narrators versus what Tom Wolfe calls "pale-beige narrators" ("New Journalism," 16), and (6) mimesis ("showing") versus diegesis ("telling").[5] The most fundamental of these techniques are those that invite the reader to relocate to the inner circle of the narrative action by dissociating the reference of the deictic elements of language, such as adverbs, tense, and pronouns, from the speech situation (i.e., the narrator's spatio-temporal location) and reassigning it from the perspective of a participant in the narrated scene. Let us consider two ways to redirect reference toward the narrative window: adverbial deictic shift and present tense.

Adverbial Deictic Shift

Literary semantics has described three ways of reporting the speech or thought of characters: direct discourse (DD) (Eveline thought, "How can I ever leave my family?"), indirect discourse (ID) (Eveline thought that she would never be able to leave her family), and the predominantly fictional free indirect discourse (FID) (How could she ever leave her family, thought Eveline). One of the syntactic trademarks of FID is the combination of a past-tense, third-person narration with the adverbials *here, now, today, tomorrow,* rather than the expected *there, then, this day, the next day.* While the reference of the spatial and temporal shifters forces on the reader the perspective of the characters, verb tense and pronouns remain assigned from the point of view of the narrative act: "Even *now,* though she *was* over nineteen, she sometimes felt herself in danger of her father's violence" ("Eveline," 38, italics mine). Or: "If she *went, to-morrow* she would be on the sea with Frank, steaming towards Buenos Ayres" (40). David Zubin and Lynne Hewitt describe the effect as follows: "The teller seems to fade into the background, and the storyworld, containing its own deictic center, comes to the fore. This is accomplished by decoupling the linguistic marking of deixis from the speech situation, and reorienting it to the major characters, the locations, and a fictive present time of the storyworld itself" ("Deictic Center," 131).

The contrast between DD, ID, and FID has been analyzed almost to the point of saturation, but nobody to my knowledge has addressed the issue of their comparative immersive power. The least immersive is clearly ID, not only because it ascribes the reference of all deictics from the point of view of the narrator but also because of its lack of mimetic properties. While ID paraphrases the quoted discourse in the narrator's vernacular, FID mimics the voice of the quoted character, and DD offers a perfect replica. DD would seem to be the most immersive of the three modes of reporting, but I would like to make a point in favor of FID. In DD, all deictics refer to a center of consciousness located on the scene, but the attributing expression ("Eveline thought") restores the perspective of the narrator and creates a movement of in-and-out between the narrative window and the larger textual world. In FID, by contrast, the reported discourse blends smoothly with the attributing phrase as well as with the rest of the narration because it maintains referential continuity on the level of the most visible and frequent deictic elements, those of tense and person. I leave it to the reader to decide what is more immersive: the form of expression that gives us a complete but temporary relocation to the narrative scene and jogs us in and out of this focal point, or the one that maintains a constant position halfway between the narrator's and the character's spatio-temporal location.

Present Tense

The verbal inflections known grammatically as tense encode many ideas, not all of which are related to time. The present tense, in English, is used for timeless statements ("Two plus two *equals* four"), for habitual, iterative events ("I *run* twenty miles per week"), for future events ("Next time I *go* shopping I will get you some snacks"), for past ones ("There were these teenagers in the park, and I walked past them, and this girl *starts* screaming at me"), and occasionally to express the (near) coincidence of an event with the time of its verbal description ("I *am* tired" or "The Babe *hits* the ball; she *is* going going gone; home run!"). In conversational storytelling and medieval epics, the so-called historical present is used in alternation with the past to channel the attention of the audience toward certain events and create a profile of mounting and declining tension (Fleischman, *Tense*, 77). The peaks of this profile correspond to events that are intensely present to the reader's imagination, and the valleys to events that are blurred in the reader's mind. Events reported in the present correspond to the peaks. The effect could work the other way around if the present were the standard narrative tense and the past the marked one, but there are good semantic and pragmatic reasons why narratives are usually told in the past: you can only make a story when the events are in the book. Moreover, as the tense of presence, the present is inherently more immersive than the past.

The effect of the contrast is skillfully exploited in two passages from Marguerite Duras's *L'Amant (The Lover)*: "Little brother died in December 1942, under Japanese occupation. I had left Saigon after my second baccalaureate, in 1931. He wrote me only once in ten years. . . . When he dies it is a gloomy day. I believe it is spring, it is April. Somebody calls me on the phone. Nothing, they don't say anything else, he was found dead, on the floor in his room" (71 and 99; my translation). Both of these passages—separated by nearly thirty pages—narrate the same event. In the first, the reader is merely informed of the death of the brother; in the second, she shares the narrator's experience of the atmosphere of the day, the breaking of the news, the tragically banal circumstances of the death. Though the narrative use of the present does not literally imply simultaneity between the occurrence of the events and the speech act of their report, as it does in "real-time narration" (such as sports broadcasts and conversations between pilots and control towers), it owes much of its expressive power to the lingering association of the tense with the idea of co-occurrence. We do not naturalize the speech situation of *The Lover* as one in which the narrator tells about her brother's death at the same time she learns about it but as a prenarrative state of consciousness. The present sends us to a moment when the narrator knows nothing more than what she hears on the phone, a moment in which she is unable to rationalize the event or even perhaps to realize the finality of its occurrence. As it creates the simulacrum of a real-time "life" (rather than speech) situation, the shift from past to present pulls the reader from the *now* of the storytelling act to the *now* of the storyworld and completes the deictic shift toward the narrative window.

Many contemporary texts exploit this pseudo-immediacy of the first-person, present-tense report to convey the experience of being swept by the flux of life, overwhelmed by unpredictable waves of events and sensations. Through its insistent use of the narrative present, contemporary narrative casts a vote of nonconfidence in the authenticity of the rational activity of retrospectively emplotting one's destiny;[6] truth, it tries to tell us, lies in the immediacy of experience, not in the artificial form imposed on one's life by narrative activity. Yet if the present enjoys an immersive edge over the past, this edge becomes considerably duller when the present invades the whole text and becomes the standard narrative tense. Continuous presence becomes habit, habit leads to invisibility, and invisibility is as good as absence. For immersion to retain its intensity, it needs a contrast of narrative modes, a constantly renegotiated distance from the narrative scene, a profile made of peaks and valleys.

The variability of the distance between the reader's implicit position and the narrated events suggests that narrative phenomenology involves not just one but two acts of recentering, one logical and the other imaginative.[7] The first—described in the previous chapter as the constitutive gesture of fictionality—sends the reader from the real world to the nonactual possible world created by the text. The second, an option available in principle to both fiction and nonfiction, though vastly more developed in the former, relocates the reader from the periphery to the heart of the storyworld and from the time of narration to the time of the narrated. This experience of being transported onto the narrative scene is so intense and demanding on the imagination that it cannot be sustained for a very long time; an important aspect of narrative art consists, therefore, of varying the distance, just as a sophisticated movie will vary the focal length of the camera lens.

Temporal Immersion

The phenomenological basis of temporal immersion is a "lived" or "human" experience of time, as opposed to what may be called "objective" or "clock" time. If clocks had a philosophical mind, they would describe time as a mechanical, meaningless, nonteleological succession of self-contained moments. Human time, by contrast, is a quasi-musical experience in which the present is not simply a moving point but a moving window that encompasses memories of the past and premonitions of the future. It is because the preceding notes survive in the present one and because the present note adumbrates its successor that we perceive melodic lines; similarly, it is because past events cast a shadow on the future and restrict the range of what can happen next that we perceive narrative lines and experience temporal immersion. Generally speaking, temporal immersion is the reader's involvement in the process by which the progression of narrative time distills the field of the potential, selecting one branch as the actual, confining the others to the realm of the forever virtual, or counterfactual, and as a result of this selection continually generates new ranges of virtualities. The passing of time matters to the reader because it is not a mere accumulation of time particles but a process of disclosure.

The art of creating interest in a story requires an efficient interleaving of two temporal dynamics: the dynamics inherent to the temporal unfolding of the told events, of the causal chain that constitutes story, and the dynamics inherent to the telling, to the revealing or withdrawing of the information that enables readers to apprehend the logic of the story. Meir Sternberg, who, unlike most scholars, conceives narrative in terms of the telling rather than in terms of the told, views this interplay as constitutive of narrative and narrativity: "I define *narrativity* as the play of suspense/curiosity/surprise between represented and communicative time (in whatever combination, whatever medium, whatever manifest or latent form). Along the same functional lines, I define *narrative* as a discourse where such play dominates: narrativity . . . then ascends from a possibly marginal or secondary role . . . to the status of regulating principle, first among the priorities of telling/reading" ("Telling in Time," 529). Though not everybody will endorse such an aesthetically based definition (it excludes from narrative those acts of storytelling that deliberately avoid the three effects, such as courtroom depositions), Sternberg's formulation remains important for its conception of narrative *art* as a discourse's ability to play with the reader's expectations, thereby creating what I call temporal immersion.

While all three effects depend on a timely release of information, surprise differs from the other two in that it is punctual rather than durative. Alfred Hitchcock, the master of cinematic suspense, explains the distinction between surprise and suspense as follows:

> We are now having a very innocent little chat. Let us suppose that there is a bomb underneath this table between us. Nothing happens, and then all of a sudden, "Boom!" There is an explosion. The public is surprised, but prior to this surprise it has seen an absolutely ordinary scene, of no special consequences. Now, let us take a suspense situation. The bomb is underneath the table and the audience knows it, probably because they have seen some anarchist place it there. . . . The public is longing to warn the characters on the screen: "You shouldn't be talking about such trivial matters. There's a bomb beneath you and it's about to explode!" In the first case we have given the public fifteen seconds of surprise at the moment of the explosion. In the second we have provided them with fifteen minutes of suspense. The conclusion is that whenever possible, the public must be informed. (Quoted in Truffaut, *Hitchcock*, 73)

If we conceive temporal immersion as that which keeps readers turning pages or spectators speculating about what will come next, surprise is too short-lived to provide immersion. Yet if we conceive temporal immersion as play with expectations, then surprise is not completely alien to it, since it takes the formation, and then violation, of projections to create surprise. Interpreters would not experience surprise if they weren't able to anticipate a future development. Moreover, surprise has often to do with the revelation of events that took place in the past and affect the identity of characters. For instance, in a Baroque novel, the heroes may discover that they are a brother and sister pair separated at birth, or it may be revealed that a commoner who has performed heroic deeds and with whom the princess has fallen in love is really of royal blood. When surprise comes from this kind of discovery, it sheds new light upon the past, and the interpreter will be tempted to revisit mentally the entire sequence of events, in order to locate hidden clues to the truth. We can consider this reevaluation a form of temporal immersion, since it turns the interpreter's attention toward an extended period of time, rather than being limited to a single moment.

Sternberg's discussion of narrative effects is based on three categories—suspense, curiosity, and surprise—that he places on the same level. But while surprise is clearly different from suspense, as Hitchcock convincingly shows, suspense and curiosity are less sharply opposed because both rely on the interpreter's desire to know. Here I treat curiosity as a subspecies of suspense, namely, as a desire to know focused on past events rather than on the future, as in the classical kind illustrated by Hitchcocks's example.

Suspense is not only a narrative effect, it is also an experience provided by life itself. Spectator sports such as football or baseball have a lot to teach us about this kind of experience:

- The enjoyment of the spectator is due to the fact that he roots for one of the teams and sees one outcome as vastly preferable to the other.
- Spectators participate in the action through the activity known as "armchair quarterbacking": they imagine scenarios for the action to come and make strategic decisions for the participants. This activity is made possible by the rigidity of the rules that determine the range of the possible.
- Suspense increases as the range of possibilities decreases. It is never greater than in the ninth inning or the last two minutes of the game, when the teams are running out of resources and options are reduced to sharply profiled alternatives: score now and stay alive, or fail to do so and lose the game. At the height of suspense, the ticking of the clock (if the game is limited by time) becomes strategically as important as the actions of the players. When this happens, the spectator reaches a state of complete temporal immersion.

In narrative suspense, similarly, we find the following features:

- Dramatic tension is usually correlated to the reader's interest in the hero's fate. The prototypical suspense situation occurs when a character is in danger and the reader hopes for a favorable outcome.
- Suspense is dependent on the construction of virtual scripts and events. Though it is tied to uncertainty, it must present what Noël Carroll has called "a structured horizon of anticipation" ("Paradox," 75). This horizon is given shape by potentialities that trace visible roads into the future,

such as the processes currently under way, the desires of characters, the goals to which they are committed, and the plans under execution.

- The intensity of suspense is inversely proportional to the range of possibilities. At the beginning of a story, everything can happen, and the forking paths into the future are too numerous to contemplate. The future begins to take shape when a problem arises and confronts the hero with a limited number of possible lines of action. When a line is chosen, the spectrum of possible developments is reduced to the dichotomy of one branch leading to success and another ending in failure, a polarization that marks the beginning of the climax in the action.

While the reader's curiosity in a suspenseful situation always concerns the story level of narrative, this curiosity is controlled on the discourse level by the author's strategies of divulging information. The intensity of the reader's temporal immersion depends on the focus of the suspense. In decreasing order of intensity we can distinguish four types.

1. *What suspense.* This brand of suspense is typical of action movies and thrillers. It is epitomized by the classic scene of Western movies in which the heroine is tied to the railroad tracks while a train is approaching. A memorable literary example is found in the last scene of James Joyce's "Eveline": Will the heroine follow her lover and board the ship that will take her to a new life in Argentina, or will she remain on shore, tied to the past, to Dublin, and to a joyless life of sacrifice to her family? The focus of attention in this type of suspense is the imminent resolution of a binary alternative: Will good or bad happen to the heroine? As Carroll has observed, this concern presupposes an emotional involvement in the fate of a character and a strong desire for an outcome favorable to the good guy. Since the central question is "What will happen next?" this type of suspense is favored by an order of presentation that runs parallel to the chronology of the underlying events (Brewer, "Nature of Narrative Suspense," 113). In order to facilitate the anticipation that creates suspense, the text may allow the reader to know more than the character—for instance, by describing how another character sets a trap while the hero is absent from the scene—but it typically does so without disturbing the chronological order. Through the parallelism of discourse structure and event structure, the reader or spectator lives the development of the action moment by moment and shares the perspective of the character whose fate is being played out.

2. *How (why) suspense* (Sternberg's curiosity). This is the suspense involved in finding out how Ulysses came to be held prisoner by Calypso or why the rich de

Lanty family treats a mysterious old man with such reverence in Balzac's *Sarrasine.* This type of suspense adopts the format of the enigma and is produced by what Barthes, in *S/Z,* calls the hermeneutic code. The outcome is given in advance through a phenomenon known as prolepsis (or cataphora), and the focus of attention is not the future but the prehistory of a certain state. The reader's experience of time is prospective, but this prospection is oriented and therefore teleological. While *what* suspense presupposes a choice between two branches leading in opposite directions, *how* suspense involves multiple possibilities converging toward the same point. Since the fate of the hero is known from the very beginning, involvement is not a matter of wishing for a favorable outcome but a matter of curiosity about the solution to a problem. Once the narrative has jumped backward in time and started moving forward toward the known goal, however, the reader may be caught in *what* suspense on the level of the individual episodes. Conversely, a narrative of the *what* type may lead to *how* suspense when the reader's knowledge of generic conventions leaves little doubt about the outcome. It is usually pretty clear to the spectators of Western movies that the heroine tied to the railroad tracks will be saved, but suspense can be maintained by shifting focus from the probability to the circumstances of the rescue.

3. *Who suspense.* This is the suspense of the whodunit, the effect commonly associated with murder mysteries. The reader's interest in the outcome is even more purely epistemic than in the preceding type. Just as there is no sadness for the victim and no moral revulsion for the crime, there is little or no emotional investment in the fate of the characters because all that matters is the intellectual satisfaction of solving the problem. In contrast to the previous type, which leaves open an indeterminate number of paths toward the goal, *who* suspense limits the number of solutions to the number of suspects. This greater structuration of the virtual field enables the author to withhold facts from the reader, such as the motivation of the detective during individual episodes of the investigation, without creating a sense of blind progress—the situation most hostile to suspense. The action unfolds on two temporal planes, that of the murder and that of its investigation, but the reader is never really caught in the time of the murder sequence. The past is revealed to him piece by piece, like a jigsaw puzzle, by the actions of the detective, and since the pieces do not fall into place in chronological order its experience is more that of a spatial picture, a jigsaw puzzle, than of a linear sequence. If mystery stories allow any kind of temporal immersion, it concerns the investigation sequence, and this immersion is tied to the dynamics of disclosure rather than to the unfolding of human destiny.

4. *Metasuspense,* or critical involvement with the story as verbal artifact. In metasuspense the focus of the reader's concern is to find out not what happens next in the textual world but how the author is going to tie all the strands together and give the text proper narrative form. An example of such suspense is the short story "The Assignation" by Edgar Allan Poe. The first scene of the fifteen-page text describes how a stranger rescues a child who has fallen into the canal in Venice. The mother of the child, a beautiful noblewoman, murmurs a few ominous words to the stranger: "Thou hast conquered—one hour after sunrise—we shall meet—so let it be" (197). In the next scene the narrator visits the stranger in his palace in Venice, and they engage in a lengthy (ten-page) discussion about art. On the last page of the story, they are still talking about art, and the first time I read it, I started to wonder, "How is Poe going to turn this into a story with only half a page of text left?" The author pulls the trick by having the stranger drink a glass of wine and drop dead while a messenger enters the room screaming "My mistress!—my mistress!—poisoned!" (207), thus revealing that the planned meeting was a suicide pact. In my anxious watching of the diminishing textual resources through which the suspense might be resolved, I paid more attention to the dynamics of storytelling than to the ticking away of story-time. Because it involves a point of view external to the textual world—a relation of author to reader—this last type of suspense does not properly belong to the poetics of immersion but rather involves a distanced, critical relation to the storyworld and to the author's performance. I mention it here only for its oppositional value with respect to the other three categories.

For the student of immersion, one of the most intriguing aspects of narrative suspense is what Richard Gerrig has called its resiliency, or anomalous experience. Many researchers have been puzzled by the fact that people can be caught in suspense after multiple readings. When I watched the movie *Apollo 13,* I knew that the crippled spaceship and its crew would make it back safely to Earth, but despite my certainty that everything would turn out for the best, I experienced almost unbearable tension during the scene of the return to Earth. Through the representation of a live TV broadcast, the audience is informed that the spaceship has ten seconds to enter the atmosphere; if it does not appear on the screen before the time is up, this will mean that it has disintegrated in outer space. Thus given all the facts, in classic Hitchcockian fashion, I anxiously watched the clock tick away, and my anxiety grew stronger with every passing second. Relief came after twelve seconds—which seemed like hours—when a parachute appeared on the TV screen and gracefully descended toward the sea. My involvement in the action was temporal immersion in its purest form—time had become almost tangible—and I experienced it not just once but every time I watched the movie. Repeat suspense is better documented in film than in literature, but the case of children who ask for the same story over and over again, and participate in every retelling with the same intensity, indicates that the experience is not restricted to visual media. A related phenomenon is the ability of stereotyped genres to arouse suspense even though the reader knows that in the end good things always happen to the good guys.

The resiliency of suspense leads to a paradox that poses a serious challenge to its standard definition. Carroll summarizes this paradox as follows: "Conceptually, suspense entails uncertainty. Uncertainty is a necessary condition for suspense. When uncertainty is removed from a situation, suspense evaporates" ("Paradox," 72). Anomalous suspense defies the premise of this reasoning, but as William Brewer pointedly observes, if suspense were totally immune to repetition, people could spend their whole lives rereading one very suspenseful novel, much to the chagrin of publishers ("Nature of Narrative Suspense," 120). The best account of anomalous suspense should therefore be one that tolerates some degree of recidivism but predicts a loss of intensity and eventual decay. (Even children, I assume, finally get bored when they hear the same story too many times.) At the same time, however, this account should remain focused on the suspense phenomenon itself and not invoke a shift of motivation. We must therefore eliminate as a possible explanation the otherwise valid observation that multiple readings bring satisfaction because they enable us to notice different features of the text.

For Kendall Walton, the solution to the paradox of repeated suspense is encapsulated in the make-believe theory of fiction. If the reader of fiction can pretend that the asserted facts are true when she knows that they are not and if she can derive pleasure from this act of pretense, then she can just as easily pretend on second reading that she does not know the outcome and experience all over again the thrill of disclosure: "Although Lauren knows that fictionally Jack will escape from the Giant, as she listens to still another rereading of 'Jack and the Beanstalk,' it is fictional that she does not know this—until the reading of the passage describing his escape. Fictionally she is genuinely worried about his fate and attentively follows the events as they unfold. It is fictional in her game during a given reading and telling of the story that she learns for the first time about Jack and the Giant" (*Mimesis,* 261). I find this explanation partly convincing and partly unsatisfactory. The convincing part resides in its implication of a resetting of the narrative clock. When Lauren hears the story for the second or third time, she experiences a temporal relocation that places her at the beginning of narrative time and enables her to share the prospective outlook of the hero. She lives the unfolding of his fate in the real time of a shifted present, rather than being merely informed of what happened in a fictional past. On the negative side, Walton's argument seems too dependent on a fallacious symmetry between suspending disbelief and suspending knowledge altogether. It takes little effort to "make-believe" that propositions are true even when one knows that they are not because to do so is inherent to the act of imagining. But pretending (i.e., "making it fictional") that one learns for the first time about the textual world does not erase propositional content from memory.

Repeated suspense, I would suggest, is not a matter of self-induced amnesia or of pretended ignorance but rather a matter of knowledge being superseded by a more urgent concern: the reader's emotional involvement in the fate of the hero. This suggestion predicts that of the four types of suspense described above, only the first and second (*what* and *how)* will lend themselves to repetition. We may indeed experience concern for the safe return of the crew of *Apollo 13* many times over, but we normally do not reread mystery stories because once we know who committed the crime we cannot relive the purely epistemic excitement of discovering the culprit. (If we do reread them, it is to locate the clues that we missed the first time.) By shifting the phenomenon of anomalous suspense from the epistemic to an emotional, almost existential plane, I have not resolved the paradox but have linked it to another problem. To explain how we can repeatedly experience anxiety over destinies that are already written in our memories, we must first understand how we come to invest our desires in the fate of characters who never existed.

Emotional Immersion

> [I found myself] giving more attention and tenderness to characters in books than to people in real life, not always daring to admit how much I loved them . . . those people, for whom I had panted and sobbed, and whom, at the close of the book, I would never see again, and no longer know anything about. . . . I would have wanted so much for these books to continue, and if that were impossible, to have other information on all those characters, to learn now something about their lives, to devote mine to things that might not be entirely foreign to the love they had inspired in me and whose object I was suddenly missing . . . beings who tomorrow would be but names on a forgotten page, in a book having no connection with life. —*Marcel Proust*

Ever since Aristotle defined the effect of tragedy as catharsis, or purification through terror and pity, it has been taken for granted that literary fictions can elicit in the reader emotional reactions such as empathy, sadness, relief, laughter, admiration, spite, fear, and even sexual arousal. The tears amply shed by Romantic or Victorian audiences over the fates of Young Werther or Little Nell give the testimony of the body in support of this assumption. Emotional participation in the fate of imaginary characters was accepted as a natural response to literature until textualist approaches overtook realist paradigms and dissolved the human essence of characters into actantial roles or aggregates of textually specified features (or "semes"). In the heyday of structuralism and deconstruction, it became heretical even to mention the phenomenon of emotional response. Who would relate emotionally to the matrix [+ turbulence, + artistic gift, + independence, + excess, + femininity (though + male)], as Barthes describes in *S/Z* the entity referred to by the name Sarrasine in Balzac's eponymous story?[8] Regarded as unproblematic by some and as illegitimate by others, the question of emotional response did not emerge as a topic worthy of serious theoretical consideration until the advent of theories, such as those described in chapter 3, that regard characters as pseudo human beings rather than as collections of semantic features.[9]

Since the first edition of this book was published, research and theories concerned with emotion have exploded. In psychology and neuroscience, emotion (or affect) has gone from being opposed to cognition (which was limited to rational reasoning), to being regarded as an autonomous dimension of cognition, all the way to being viewed as playing a critical role in high-level cognition (Keen, *Empathy*, 27). Cognitive literary studies have tried to keep up with this trend by emphasizing empathy, caring for characters, and the therapeutic role of fiction in helping us understand our own emotions (Oatley, *Such Stuff*, 126), but its treatment of emotions has remained largely generic. We hear a lot about the emotional impact of fiction but a lot less about individual types of emotions and how they may affect readers or spectators. While it is true that fiction can represent the entire spectrum of human emotions and activate in certain readers memories of having felt them, one should distinguish represented emotions (i.e., the emotions of characters) from the emotions experienced by readers. Among the emotions that can be represented but do not seem to affect readers are self-centered feelings such as jealousy, envy, and wrath. People who have had doubts about the faithfulness of a loved one may empathize more deeply than others with the plight of Othello or of Proust's narrator (who speculates obsessively about the sexual preference of his lover Albertine); but these people will not feel jealous themselves when reading a text about jealousy. (They may be jealous of the talent of the author, but this is another question.) I consider the following types of emotion conducive to immersion. (Note that a given emotion, as identified by a name, can occupy several categories.) They are listed here in increasing order of affective intensity.

1. Subjective reactions to characters and judgments of their behavior. These include primarily like and dislike but also admiration, contempt, pity, amusement, Schadenfreude (when bad characters get their comeuppance), and exasperation (when good things happen to bad characters).
2. Empathetic emotions, that is, emotions felt not for oneself but for others. Suzanne Keen defines empathy as "a spontaneous, vicarious sharing of affect" and as "feeling what we believe to be the emotions of others" ("A Theory of Narrative Empathy," 208). Empathetic emotions fall into the two broad categories of feeling sad/happy for characters, though they may be described in finer shades, such as pity, grief, and relief.
3. Emotions felt for oneself, not for others, such as fear, horror, disgust, and sexual arousal.

Type 1 is not particularly problematic because it involves distanced evaluation rather than identification with the characters. Feeling admiration or contempt for characters can be explained by counterfactual reasoning, such as, "if an *x* behaved like *p*, I would admire *x*," and it does not consequently imply belief in the existence of *x*. Yet the counterfactual explanation does not properly account for the gut feeling of liking or disliking characters. Nor does it work very well for empathetic feelings of type 2. While it would be conceivable to explain pity for a character by thinking "if an *x* underwent experiences *p* in the real world, I would pity *x*," this reasoning implies a distance that makes it incompatible with any kind of immersive sharing of the character's emotions. Yet an explanation of empathy based on a full emotional identification with the characters is not satisfactory either because it would in many cases make the reading of fiction into an unpleasant experience. The strongest feelings of empathy occur when bad things happen to likable characters, but readers do not experience grief for fictional characters in the same way they experience it for real people. They do, and do not, feel sorry for them. How can one get out of this impasse?

Gregory Currie states the paradox of empathy for fictional characters in terms of the following rules:

I. We have emotions concerning fictional situations.
II. To have an emotion concerning a situation, we must believe the propositions that describe this situation.

III. We do not believe the propositions that describe the situations represented in fiction. (*Nature of Fiction,* 187)

As Currie observes, these propositions are incompatible: "The problem is to decide which one [should be removed], and what to replace it with" (187). Rejecting III would conflict with the basic tenets of semantically based theories of fiction, and rejecting II would mean that emotion does not require belief in the existence of its object—an outright denial of what Currie calls the cognitivist conception of emotion. Since both of these positions are unacceptable to Currie, the only alternative left is to reject proposition I. For Currie, emotions are nexuses of relationships between feelings (sadness, happiness), beliefs, and desires: for instance, my belief that a fire destroyed a certain forest, coupled with my desire for the maintenance of the beauty of this forest, will result in sadness. But our propositional attitudes toward fictional characters and situations cannot be beliefs and desires, since we know that they do not exist. Currie proposes to resolve the paradox of emotional reactions by distinguishing two parallel brands of experience: in real life we have beliefs and desires leading to emotions; in fiction we have make-beliefs and make-desires leading to "quasi-emotions." But what do we learn from this seemingly ad hoc sprinkling of prefixes? While the idea of "quasi-emotion" goes some way toward explaining why crying for Little Nell does not make us really sad (we enjoy our sadness), it does not seem right to account for this reaction in terms of make-believe, if by make-believe one understands a deliberate act of pretense. An actor may pretend to be sad, as required by the script of the play, but the readers of fiction do not deliberately play a role scripted by the author, even though their emotions are induced by the words of the author.

In contrast to the philosophical school's loyalty to what Currie calls the cognitivist position—rejecting proposition I in order to maintain II—psychologists are much more inclined to endorse some version of I and to question the validity of a rigid interpretation of II. Evolutionary psychologists view empathy for other people as one of those evolved abilities that enable humans to form the strong social groups necessary to their survival as a species. Once we develop empathy, it does not have to be limited to people we know nor to real people. According to Blakey Vermeule (*Why Do We Care,* preface), humans are such intensely social beings that they care not only about close friends and relatives but also about people they do not know personally, such as celebrities. This curiosity easily extends to fictional characters and even to animals or inanimate objects that are endowed by the imagination with human properties. While from a philosophical perspective there is an unbridgeable ontological gap between mourning for Princess Diana and mourning for Little Nell, from a pragmatic perspective the gap is not that great, since for most people Princess Diana inhabited a realm as remote and unreachable as a fictional world. Now, if the dependency of emotion on proposition II can be loosened, the intensity of emotion can be loosened accordingly. The more remote a realm a person inhabits, the less intense the emotions felt for that person.

Psychologist Richard Gerrig has another explanation for the phenomenon of emotional responses for nonexisting persons. He attributes it to an independently documented principle that he calls the "nonpenetration of belief into emotional experience" (*Experiencing Narrative Worlds,* 181). This principle has been invoked by psychologists to explain the case of phobias: "The beliefs that dominate the appraisal of snakes or spiders at a distance are not able to penetrate the mental processes that produce the extreme emotional responses when the subjects are at hand. Such clinical instances demonstrate a clear capacity for individuals to experience strong emotions that are not ameliorated by beliefs" (183). While the purely cognitive approach adopted by Currie regards genuine beliefs as always present to the appreciator's mind, even in the midst of make-believe, the psychological approach suggests that the availability of beliefs is subject to variations. In this interpretation, emotional responses to fictional characters are due to the temporary backgrounding of proposition II, a backgrounding that can be regarded as a necessary condition of emotional immersion.

Of all the types of emotions listed above, the third—self-centered emotions—is the most difficult to explain because as a phenomenon that affects the person of the reader (or her alter ego in the fictional world), it requires the deepest immersion. While we experience empathy for people we hear about, we can only experience fear, disgust, or sexual arousal if we imagine ourselves as directly facing the objects that inspire such reactions. Self-centered emotions are inextricably tied to a sense of presence, and they are therefore much more predominant in visual media than in literature. To explore the power of fiction to inspire self-centered emotions, I asked the members of an electronic discussion group on narrative—all academics and therefore professional readers—to give examples of fearful reading or movie-watching experiences and to try to explain the feeling in terms of the relations between the self, the real world, and the fictional world. Here is a sampling of the responses:

1. "Last summer, I took my then-three-year-old son to see a children's performance of 'Little Red Riding Hood,' a story that he had loved until then and had never feared. However, when he was confronted with 'real' people representing the

characters, he was confused and frightened. The climax (for us) came when the wolf stepped off the stage as he was singing and sashayed down the central aisle; this proximity was simply too much for my son to handle, and we were forced to leave without ever finding out whether the wolf was truly bad or not! My son does not seem to have that problem with videos or books; it was the 'realness' of the actors/actresses that created confusion for him, as it forced him to reconsider the usual lines between 'fact' and 'fiction.'" (Janet Galligani Casey)

2. "When I was in kindergarten, a foolish parent took all the children at her child's birthday party to a horror movie. I had recurrent nightmares for two years. I can still remember the fear I felt imagining that the monster, which had erupted out of the mountain, was about to step on me." (Deborah Martinsen)

3. "I like to be made to cry, but I hate being scared. . . . The specific fears that I am always trying to avoid are very patently death-related (dead bodies, skeletons, buried-alive scenarios, etc.); they seem to be very basic fears and denials of mortality." (Antje Schaum Anderson)

4. "[I must also mention] my experience of violent vertiginous terror while seated in the Boston IMAX theater watching their introductory promo movie, which is shot from a small-plane view and features a lot of darting up and down between tall buildings, sudden dips toward bridges, etc." (Caroline Webb)

5. "What really terrifies me is always something that at first seems to be far from everyday life but . . . soon turns out to be something that could happen to everyone, myself included." (Edina Szalay)

6. "When reading novels . . . the fear I experience is more often anxiety that the characters will jeopardize their futures. . . . My fear could best be described as protective to the characters, but it has also quite definitely been fear for myself." (Caroline Webb)

7. "I think the key [to fear] is identification—the stronger the identification with a character or situation, the more the borders between the fiction and reality, self and other, tend to collapse." (Charlotte Berkowitz)

8. "As a teenager I remember pleading with the screen while watching Zeffirelli's *Romeo and Juliet*—oh please, don't follow the script." (Deborah Martinsen)

9. "Although I haven't read the novel for several years, every time I reread *Tess of the d'Urbervilles* I was hoping it would turn out differently, that Angel Clare wouldn't be such a jerk. I knew the plot, but every time I reached one of the several crucial scenes, I would find myself hoping it would turn out differently this time." (Laura Beard)

10. "In *Native Son* [by Richard Wright], there is a moment when Bigger Thomas

is trapped in Mary Dalton's room by her mother (who is blind). The fear he feels leads him to accidentally smother Mary. The scene evokes a powerful sense of fear in readers (myself included). I wonder whether this scene moves us to fear by analogizing to the idea or experience of [a] similar, though not identical, sense of entrapment." (Ted Mason)

This last post awakened in me the following literary reminiscence:

11. One of the passages in fiction that I find the most scary is a scene from Emile Zola's *Thérèse Raquin.* It also plays on the idea of entrapment. Thérèse and her lover have murdered Thérèse's husband and now live together with the mother of the victim. The mother-in-law comes to realize who has murdered her son, but she suffers a stroke and is unable to speak or move. The end of the novel describes the deteriorating relationship and growing torture of the two lovers as they suffer the unbearable presence of the silent, hostile gaze of that woman trapped in her own body. I read it as a teenager, and I still shudder when I replay it in imagination.

Many of these examples can be explained as a collapse of the distinction between the real and the fictional world. In 1 and 2, ontological boundaries are perceived as permeable, and the real-world self believes itself to be exposed to the dangers of the fictional world. As the various examples indicate, this experience affects mostly children. Psychologists have observed that while children learn at a very early age (before four) the difference between make-believe and reality, the consequences of the distinction are slower to sink in. Young children "still remain unsure of the rules that govern transformation between [the realms of fantasy and reality]" (P. L. E. Harris et al., quoted in Gerrig, *Experiencing Narrative Worlds,* 193). In example 3, the reader's or spectator's fear is due to the repulsive character of the object represented. More prominent in visual media and closely related to the sexual arousal of the consumer of pornography, this type of fear is dependent on the real-life (or illusionist) quality of the representation, on its ability to erase the distinction between image and reality. The reaction described in example 4 is due to a property of the image that creates a very physical experience of dizziness, vertigo, and loss of footing in the real world. When a movie camera is mounted on a moving object—roller coaster, racing car, or airplane—the spectator is projected as a passenger of the vehicle and becomes a potential victim of a crash. The purely imaginative possibility of smashing into the ground creates a sensation closely related to the thrill of riding a roller coaster, an adrenaline rush that represents both fear and a very physical form of pleasure. In examples 5 and 6, the fictional world is apprehended as a possible world in the literal sense of the term: if these events can happen in the textual world, thinks the reader, they can happen in reality, and if they can happen in reality, they can happen to me.[10] This cause of fear is the most rational of the list, but in 5 it is linked to more personal, subconscious anxieties, and in 6 it combines with the phenomenon of empathy for characters. Fear directed toward others is also illustrated in examples 7, 8, and 9, leading in 8 and 9 to anomalous suspense. In examples 10 and 11, finally, as also in 3 and 5, the text awakens personal phobias (e.g., a sense of entrapment) and existential anguish (fear of death, darkness, the Other, or helplessness).

For philosophically oriented theorists of fiction, the truly problematic cases of self-centered emotion are those that do not involve a blurring of ontological boundaries. Logicians tend to dismiss examples 1 and 2 as experiences due to immaturity—though cognitive psychologists might see in them important documents of the functioning of the imagination—and regard examples 4 to 6 as instances of tricking the senses. But what about the other cases? In a discussion of the related phenomenon of anomalous suspense, which also involves a backgrounding of knowledge, philosopher Noël Carroll offers a logical explanation of participatory responses to fiction fully compatible with the experience of a mature reader aware of the nature of fictionality. He bases his analysis on a distinction between two modes of considering (and offering) propositions. In the mode he calls belief, propositions are held before the mind as asserted; in the mode he calls thought, propositions are contemplated but not asserted. This is the mode of fictional representation: "The author, in presenting his or her novel as a fiction, in effect, says to readers 'hold these propositions before your mind unasserted'—that is, 'suppose p,' or 'entertain p unasserted,' or 'contemplate p unasserted'" ("Paradox," 85). (Alternatively, the author could be saying, "Contemplate these propositions as asserted by the narrator.") Emotional responses, however, are not sensitive to the distinction between asserted and unasserted propositions: "One can engender emotional states by holding propositions before the mind unasserted. Thus, when I stand near the edge of the roof of a high building and I entertain the thought that I am losing my footing, I can make myself feel a surge of vertigo. I need not believe that I am losing my footing; I merely entertain the thought" (ibid.).

Carroll regards this capacity of the human mind to be emotionally affected by the contemplation of purely imaginary states of affairs as an evolutionary asset that works toward the preservation of the species. As Walton also argues ("Spelunking"), it is on the playground of mental simulation, where emotions happen but do not count, that we learn about our own feelings and prepare ourselves for the trials of life. In the case of fictional suspense, both first time and repeated, the reader contemplates the unasserted proposition that the hero is in danger, and the intensity of this contemplation generates the emotional response of anxiety. What Carroll is saying here is that to the simulating mind, it does not matter whether the envisioned state of affairs is true or false and its development known or unknown because simulation makes it temporarily true and present, and from the point of view of the present, the future has not happened. I can, for instance, relive many times in imagination an accident in which I was almost killed and experience the same terror every time. The power of the imagination to make situations present to the mind is dramatically exemplified by the sexual arousal of consumers of pornographic literature. Readers of pornographic novels can be sexually stimulated by the depiction of a sex act, even though they know that it was all made up, because what arouses them is the vividness of their mental representation of the scene, which itself is a response to the vividness of the textual description. In the case of other types of fiction, similarly, the presence of the scene to the mind creates emotional responses that lead to an adrenaline rush, a sensation we enjoy in certain doses, but the backgrounded knowledge that it was all made up and does not affect a human life holds the dosage within the limits of pleasure.

III: The Poetics of Interactivity

FIVE

The Text as World versus the Text as Game

"There's glory for you!" [said Humpty Dumpty.]

"I don't know what you mean by 'glory,'" Alice said.

Humpty Dumpty smiled contemptuously. "Of course you don't—till I tell you. I meant 'there's a nice knock-down argument for you!'"

"But 'glory' doesn't mean 'a nice knock-down argument,'" Alice objected.

"When I use a word," Humpty Dumpty said in a rather scornful tone, "it means just what I choose it to mean—neither more nor less."

"The question is," said Alice, "whether you can make words mean so many different things."

"The question is," said Humpty Dumpty, "which is to be master—that's all."

—*Lewis Carroll*

As we have seen in the introduction, opposition to immersion—in both its literary and technological versions—runs rampant in contemporary criticism. The experience requires a transparency of the medium that makes it incompatible with self-reflectivity, one of the favorite effects of postmodernism. Sven Birkerts captures in no uncertain terms the phenomenon of the disappearance of signs: "We project ourselves at the word and pass through it as through a turnstile. . . . A reader in the full flush of absorption will not be aware of turning words into mental entities" (*Gutenberg Elegies,* 81). In an age that regards signs as the substance of all realities, this traversal of language is a semiotic crime: whatever "freedom from signs" the mind can reach is achieved not through their disappearance but through the awareness of their omnipresence and through the recognition of their conventional and arbitrary character. Signs must be made visible for their role in the construction of reality to be recognized. The idea of transparency tends to be interpreted as a denial of the importance of the medium in what can be expressed and represented. If the medium is transparent, so the argument goes, the medium does not matter.

On the contrary, I would like to argue that the disappearing act of the transparent medium is not a lack of autonomous properties but a hard-won and significant property that plays a crucial role in shaping the experience of the appreciator. It matters crucially that some media, and some representations within a given medium, achieve greater transparency than others. The traversal of signs is to be deplored only when it causes signs to vanish permanently, when immersion is so deep that it precludes a return to the surface. If there is such a thing as a "truth universally acknowledged" by literary theorists, this truth is that attention to the rhetorical devices through which a world emerges out of words is an essential aspect of aesthetic appreciation. To restore contact with the surface, we need an alternative to the metaphor of the text as world that complements, rather than invalidates, the poetics of immersion. This chapter is devoted to the most prominent of the possible candidates.

The Text as Game

Textual worlds reached their greatest expansion and maximal consistency in the novels of Balzac, Dickens, Tolstoy, Dostoevsky, and Proust. Then they began to collapse under their own weight. In the second half of the twentieth century, a process of shrinking, fissuring, splitting, and multiplying worlds within a larger textual universe reduced big worlds to little worlds or dismantled them into heterogeneous fragments. Their scattered remnants could no longer build a coherent imaginary space and time, but they provided the perfect material for play. This is how the metaphor of the text as world came to be supplanted by the metaphor of the text as game, not only as the dominant aesthetic guideline for the production of texts but also as a critical paradigm that promoted a rereading of the texts of the past.

The concept of game is one of the most prominent in twentieth-century thought. The list of its manifestations covers such various projects and phenomena as Johan Huizinga's groundbreaking study *Homo Ludens*, Roger Caillois's taxonomy of games, Ludwig Wittgenstein's notion of language games, Hans Vaihinger's philosophy of "as if," John von Neumann's development of the mathematical field of game theory, Jaakko Hintikka's "game-theoretical semantics," Jacques Derrida's doctrine of the play of signs, Vladimir Nabokov's and Italo Calvino's fascination with combinatorics, the Oulipo movement's promotion of games as literary structures, pop psychologist Eric Berne's description of human behavior as "games people play," all the way to the advent of computer culture and electronic games, and the rising prominence of gambling and competitive sports in leisure culture.

Among postmodernists, game enjoys special favor because it exemplifies the elusive character of the signified and the slippery nature of language. In one of the most famous pronouncements in the philosophy of language, Wittgenstein describes the concept of game as a "family resemblance" notion that cannot be held together by a stable core of common features.

> Consider for example the proceedings that we call "games." I mean board-games, card-games, ball-games, Olympic games, and so on. What is common of them all?—Don't say: "There *must* be something common, or they would not be called 'games'"—but *look* and *see* whether there is anything that is common to *all*, but similarities, relationships, and a whole series of them at that. . . . [We] see a complicated network of similarities overlapping and crisscrossing: sometimes overall similarities, sometimes similarities of detail.
>
> I can think of no better expression to characterize these similarities than "family-resemblance"; for various resemblances between members of a family: build, feature, color of eyes, gait, temperament, etc. etc. overlap and crisscross in the same way.—And I shall say: "games" form a family. (*Philosophical Investigations*, segment 1–32)

What constitutes a family, however, is not resemblance but kinship relations. The set of games may be fuzzy, which means that there is no set of necessary *and* sufficient conditions for an activity to be covered by the word *game,* but there still may be a set of necessary conditions surrounded by nonnuclear but reasonably typical properties whose number determines the degree of "gameness" of an activity.

One feature that comes to mind as a necessary condition is pleasure; games are freely played and played for their own sake. If we focus on institutionalized games, as opposed to free play—a distinction made in English by the contrast *game/play* but expressed in neither the French *jeu/jouer* nor the German *spiel/spielen*—we can add a second basic feature: games are constituted by rules, and these rules, as Huizinga observes, are "absolutely binding and allow no doubt" (*Homo Ludens,* 11). To refer to rule-governed games, Caillois borrows from Plato the concept of *ludus* (behavior requiring "effort, adroitness, and ingenuity on the part of the players") and opposes it to *paidia,* a term of his own making that denotes an activity characterized by "fun, turbulence, free improvisation, and fantasy" (*Men, Play, and Games,* 13). Bernard Suits—who strongly disagrees with Wittgenstein's pronouncement regarding the elusive character of the concept of game—proposes a definition entirely based on the notion of rules: "To play a game is to engage in an activity directed towards bringing about a specific state of affairs, using only means permitted by rules, where the rules prohibit more efficient in favor of less efficient means, and where such rules are accepted just because they make possible such activity" (*Grasshopper,* 34). This definition mentions only one of the two core features, but pleasure is not alien to it, since without enjoyment there would be no point in choosing a difficult route over an easy one.

Surrounding the two candidates for nuclear properties is a cluster of widely distributed but nonbinding features. As Huizinga explains,

> Summing up the formal characteristics of play [Huizinga means here the playing of formalized games] we might call it a free activity standing quite consciously outside "ordinary" life as being "not serious," but at the same time absorbing the player intensely and utterly. It is an activity connected with no material interest, and no profit can be gained by it. It proceeds within its own proper boundaries of time and space according to fixed rules and in an orderly manner. It promotes the formation of social groupings which tend to surround themselves with secrecy and to stress their difference from the common world by disguise and other means. (*Homo Ludens,* 13)

The flexible character of the concept explains why there is no specific game that serves as prototype for the textual analogy. Literary critics invoke game and play in a loose generic way, and they feel free to borrow features from a number of different games.

Another variable in the literary usage is the nature of the relation between "game" and "text." A text can be called a game literally, metonymically, or metaphorically, the latter in a narrow or a broad way. (As Suits observes ["Detective Story," 200], there is plenty of "loose talk" in the use of the term by literary critics.) The closest thing to literary games in the literal sense of the term—unless computer games are regarded as literature—are folklore or literary genres regulated by fixed formal constraints, such as rhymes, puns, anagrams, acrostics, palindromes, or the secret languages invented by children ("pig Latins"). Whether they concern the graphic, phonic, or semantic substance of words, the rules that define these genres clearly throw "unnecessary obstacles" in the way of message formation. These voluntarily chosen obstacles bracket out the utilitarian, referential function of language and turn words into toys, much in the way that children at play recycle broomsticks as horses and cardboard boxes as fairy-tale castles. In these textual games, the partners in play are language and the writer who completes the fixed pattern in the most ingenious way, but from the point of view of the receiver, the games are more a spectator sport than a participatory activity.

Another way for a text to be a game in a fairly literal sense is by offering a problem to the reader. In genres such as riddle, tongue twister, crossword puzzle, or even mystery novel, both author and reader are involved as players, and each of them performs exactly one move. The author asks a question, the reader tries to answer, and the author wins if the reader must be given the solution (Suits, "Detective Story," 204). In text-based computer games, a variant of this category, the author's one move is a global design that foresees or embeds many moves by the user. Within the text-as-problem category, however, we should distinguish those texts in which the user is supposed to find the answer by following textual directions (crossword puzzles, computer games) from a more metaphorical brand of game in which the pleasure of the reader is actually greater when he cannot solve the problem because he is more interested in a clever answer than in beating his opponent.

A last way for language to be implicated in a game situation is to participate in a competition in which the skills necessary to win involve a mastery of verbal art. In institutionalized speech events such as verbal dueling, oratory debate, and poetry contests, the production of text counts as a strategic move. A more familiar form of ludic text production is the mad-lib party game, in which funny texts are generated through the random combination of words chosen by the players. It is also in the literal game category that I would place certain uses of language made possible by the Internet, such as those collaborative literary projects in which participants perform moves by adding text to an arborescent or rhizomatic structure. A project developed by Bonnie Mitchell, *Digital Journey,* closely resembles the "telephone" children's game: "She and her students create visual or verbal 'starter images' or 'theme pages' and circulate them among other individuals nationally or internationally. Each participant alters, adds to, or comments on whatever he or she receives." At the end of the game, a dramatically changed text returns to its original sender (quoted in Gaggi, *From Text to Hypertext,* 139).

In the metonymic interpretation of game, the manipulation of the text by the reader involves an action borrowed from a standard type of game, such as throwing dice-like objects (the *I Ching*), shuffling cards (Marc Saporta's *Composition No 1,* a novel written on loose pages that can be read in any sequence), or even clicking with a mouse on the links of a hypertextual network, an activity reminiscent of opening the windows of an Advent calendar to find hidden treasures of candy or pictures. In the narrow metaphorical sense, the text is not in itself an object to be manipulated in a ludic activity but a verbal transposition of the structure of a nonverbal game. Examples of texts explicitly patterned after games abound in postmodern literature. Italo Calvino's *Castle of Crossed Destinies* simulates the card game of Tarot, the order in which the reader meets the tenants of an apartment building in Georges Perec's *La Vie mode d'emploi* is determined by the knight's tour of the chessboard, and the texts of ∈, a collection of poems by Jacques Roubaud, represent the tokens of the Japanese game of Go (each text includes spatial coordinates that specify what move it will perform when it is placed on a game board). In this narrow analogical implementation, as in the metonymic and literal interpretations, the concept of game functions as a discriminatory factor—some texts deliberately simulate games and others do not.

As the scope of the metaphor broadens, the analogical support becomes more and more tenuous. If the concept of game is to capture the essence of the literary, rather than the formal, properties of individual texts, it must be reduced to its bare essentials—a rule-governed activity, undertaken for the sake of enjoyment. The game metaphor becomes the expression of an aesthetic ideal in which rules do not function as unnecessary obstacles thrown in the player's way, as Suits defined their role, or as a tyrannical authority that must be subverted but rather as a means toward a goal. A successful game is a *global design that warrants the player's active and pleasurable participation in the game-world*—the term *world* being taken here not as the sum of imagined objects but in a nonfigurative sense, as the delimited space and time in which the game takes place, what Huizinga (10) calls the Magic Circle. At this level of generality, however, the metaphor threatens to degenerate into triviality. A good metaphor is supposed to provide an original perspective on its tenor, but we do not learn anything that we did not know before by being told that reading literary texts, like playing games, is for most people a pleasurable, nonutilitarian activity. Nor do we need to invoke the concept of game to realize that literature consists to a great extent of verbal sequences put together according to the rules of a specialized code, such as figures of speech, metric forms, standardized themes, conventional images, narrative patterns, and dramatic structures.

Beyond the superficial similarity of game rules and literary rules, however, the differences in their mode of operation are too important to ignore:[1]

1. Game rules are what John Searle (*Speech Acts,* 33–41) calls constitutive principles (they create the game by defining it), but in literary codes descriptive conventions ("regulative rules," for Searle) are much more dominant. A convention arises when a large number of authors conform

to a certain type of behavior, but it is not binding, and it describes an independently existing form of discourse.

2. Game rules must be followed strictly, while modern literature encourages creativity and the transgression of its own conventions. The legacy of Russian formalism has accustomed scholars to regard deviance from the norm as a trademark of literary language.
3. If every text creates its own rules, the reader learns the code in the process of playing. This contrasts with the standard game situation, in which the player must learn the rules before stepping onto the playing field.[2]

Another feature of games that seems inapplicable to the literary text is the specificity of the goal to be achieved. While the attainment of this goal—beating the game or the opponent—puts an end to the playing, the reading of a literary text is widely regarded as a never-ending activity. According to one of the most sacred dogmas of literary hermeneutics, a discipline whose origin in theology shouldn't be forgotten, the literary text is a holy scripture whose meaning cannot be exhausted.

If the core features of the game concept fail to provide interesting insights into the nature of the literary text, it is because the project of drawing a parallel between what is common to all games, on one hand, and to all literary textual practices, on the other, disregards the heterogeneity of the two sets under consideration. The overall analogy between the two domains is better supported by many concrete local similarities than by a few vague global parallels. A useful point of departure for the exploration of these local similarities is the typology outlined by Roger Caillois in *Men, Play, and Games.* Caillois distinguishes four types of game.

1. *Agon.* Games based on competition, such as sports (soccer, tennis), board games (chess, Go), or TV quiz games *(Jeopardy).*
2. *Alea.* Games of chance, such as roulette and lottery.
3. *Mimicry.* Games of imitation and make-believe, such as children playing house or making mud pies.
4. *Ilinx.* A protean family that involves transgression of boundaries, metamorphosis, reversal of established categories, and temporary chaos. Caillois associates *ilinx* with "those games which are based on the pursuit of vertigo and which consist of an attempt to momentarily destroy the stability of perception and inflict a kind of voluptuous panic on an otherwise lucid mind" (23). The examples that come to mind are not activities that we easily associate with games: initiation rituals, drug experiences, cross-dressing and masquerades, and the scary rides of amusement parks.

The usefulness of the first category, agon, is largely restricted to the case of the literal domain.[3] In a computer game, the purpose is clearly to win, and the way to win is to defeat enemies. When all obstacles have been overcome, all riddles solved, and the highest level conquered, the player is said to have "beaten the game." Agon is also a driving force on the thematic level of narrative texts—most, if not all plots involve some sort of competition—but the concept has little to offer on the metatextual level. Wolfgang Iser likens the act of reading to playing a game in which the reader can either win by "achieving meaning" or "maintain freeplay" (a state akin to a tie) by "keeping meaning open-ended ("Play of the Text," 252), but the text certainly does not lose when the reader wins. If maintaining free play is superior to achieving meaning, as postmodern theory tells us, the optimal situation is for the reader and the text to tie. This is hardly a competitive situation.

It is perhaps in some types of hypertext fiction that we find the best example of an antagonism between the reader and the text. Espen Aarseth suggests the idea of a quest for the plot in his discussion of Michael Joyce's *afternoon:* "We might label *[a]fternoon a reluctant narrative,* or an *antinarrative,* or a *sabotaged narrative,* terms typical of modernist poetics. But perhaps the best descriptive term for *[a]fternoon* is *game of narration*" (*Cybertext,* 94). I take this last characterization to mean that the reader of *afternoon* is motivated by the desire to unscramble and put back together the narrative body that comes to her in dismembered form. This is certainly the spirit in which I approached the text,[4] though I suspected—rightly, as it turned out—that the puzzle had no definitive solution. Jay Bolter also describes the hypertext experience in mildly agonistic terms, but winning or losing does not really matter because the player will always get another turn:

> Playfulness is a defining quality of this new medium. . . . No matter how competitive, the experience of reading in the electronic medium remains a game, rather than a combat, in the sense that it has no finality. . . . The reader may win one day and lose the next. The computer erases the program and offers the reader a fresh start—all wounds healed. . . . The impermanence of electronic literature cuts both ways: as there is no lasting success, there is also no failure that needs to last. By contrast, there is a solemnity at the center of printed literature—even comedy, romance, and satire—because of the immutability of the printed page. (*Writing Space,* 130)

One wonders, however, how the reader of hypertext can lose, even temporarily, given Bolter's earlier claim that there are no bad choices—no aporia, to use Aarseth's term—in hypertext: "[A] hypertext has no canonical order. Every path defines an equally convincing and appropriate reading" (*Writing Space*, 25). Here the labyrinth is no longer a problem to solve but a playground for the reader. This conceptualization removes any competitive spirit from the reading of hypertext and suggests an experience much closer to what will be described below as *ilinx* or as free play.

The idea of turning the text into *alea*, a game of chance, has detractors and supporters among the advocates of the game aesthetics. In what Warren Motte calls "a crusade for the maximal determination of the literary sign" (*Oulipo*, 17), a crusade whose ultimate goal is to purify literary language from the randomness of everyday speech, the members of the French literary movement Oulipo promote the use of exacting formal constraints, such as writing an entire novel without using a certain letter (*e* in Georges Perec's *La Disparition*). By restricting the choice of words to those that can fill a certain graphic or phonic pattern, however, formal constraints give free rein to chance on the semantic level. The more stringent the formal requirements, the less the meaning of the text will express a preexisting vision, and the more it will be produced by the random encounter of signifieds brought together by the similarity of the signifiers. Oulipo's aversion to chance is also undermined by its predilection for combinatorics. Italo Calvino, a marginal member of the group, defines literature as "a combinatorial game which plays on the possibilities intrinsic to its own material" (quoted in Motte, *Playtexts*, 202). The principle is put to work in Raymond Queneau's *Cent mille milliards de poèmes*, a collection of ten sonnets of fourteen lines each printed on pages cut into so many strips, so that by flipping the strips and combining the lines, the reader can obtain 10^{14} different texts. Other examples of the literary exploitation of aleatory principles include the already mentioned *Composition No 1* by Marc Saporta, the Surrealistic technique of *écriture automatique*, the practice of generating texts by computer through the random selection and combination of words from a database (the electronic poetry of John Cayley), and the reliance of literary hypertext on the often blind clicking of the reader to keep the textual machine running. If there is a common message to be read from these various attempts to exploit the creative power of chance, it is that meaning is produced by a force outside human control, a force that emanates from the substance of language.

The importance of games of mimicry for the theory of fictionality has been discussed in the previous chapters. Since these games presuppose a world, they offer a potential reconciliation of immersion and interactivity, and they transcend the aesthetic ideals that the literary theory of the past twenty years seeks to express through the game metaphor. I do not, therefore, include mimicry in my discussion of the game aesthetics.

With *ilinx,* we leave the realm of rule-governed activity and enter the domain of free play. In the history of philosophy, as Mihai Spariosu has shown (*Dionysus Reborn*), the concept of play is traditionally associated with the creative power of the imagination. Children at play do not follow established rules but invent their own in a transgression of real-world identities: "I'll be the salesperson, you'll be the buyer, this bench will be the store, and these pebbles will be candy." In literature, *ilinx* and its free play are represented by what Bakhtin calls the carnivalesque: chaotic structures, creative anarchy, parody, absurdity, heteroglossia, word invention, subversion of conventional meanings (à la Humpty Dumpty), figural displacements, puns, disruption of syntax, *mélange des genres,* misquotation, masquerade, the transgression of ontological boundaries (pictures coming to life, characters interacting with their author), the treatment of identity as a plural, changeable image—in short, the destabilization of all structures, including those created by the text itself. Rimbaud called this spirit "déréglement de tous les sens" (disruption of all meanings, directions, and sensory faculties). More than any other category in Caillois's typology, *ilinx* expresses the aesthetics, sensibility, and conception of language of the postmodern age.

The prominence of the notions of game and play in contemporary literary theory cannot be entirely explained by these analogies. A more pertinent question to ask is how postmodern thought appropriates the game metaphor as the expression of its own concerns. In a period as free from tradition and formal constraints as post-modernism—Oulipo is the exception that confirms the rule—the emergence of a form of behavior constituted by "absolutely binding" rules as metaphor for the literary text is due not to nostalgia for the time when literature had to conform to rigid models inherited from the past but to a crisis of the notion of representation. This crisis is widely credited as a result of the influence of Saussurian linguistics, more particularly to its emphasis on the arbitrary character of linguistic signs. For Saussure, arbitrariness concerns not only the relation of individual signifiers to their signified—what could be called vertical arbitrariness—but, more radically, the horizontal organization that language imposes on phonic and semantic substance as it divides into discrete elements the perception of what Saussure regards as an intrinsically undifferentiated continuum. This view of arbitrariness has fueled the postmodernist denial of the power of language to describe an external reality. Terry Eagleton explains: "If, as Saussure had argued, the relation between sign and referent was an arbitrary one, how could any 'correspondence' theory of knowledge stand? Reality was not reflected by language but produced by it: it was a particular way of carving up the world which was deeply dependent on the sign-systems we had at our command, or more precisely which had us at theirs" (*Literary Theory*, 107–8).

Recent linguistic theories have adopted a more nuanced approach toward the arbitrariness of linguistic signs and their control of the mind, taking more seriously the idea of semantic universals, investigating the metaphorical motivation of linguistic expressions, acknowledging the possibility of nonverbal thinking, and taking a critical view of the doctrine of linguistic relativism (as formulated in the Sapir-Whorf hypothesis). But for many literary theorists, the most radical pronouncements of Saussurian linguistics—such as, "In language there are only differences without positive terms" (120)—remain *the* authoritative account of the nature of language, not because of their influence on current linguistics but because they support what postmodernism wants to believe.

The idea of linguistic relativism is a seductive plaything for thinkers who conceive thought itself as play. In the paradigm that currently dominates literary studies, if literature is a game, it is because language itself is one; and if language is a game, it is because its rules form a self-enclosed system that determines, rather than reflects, our experience of reality. This autonomy of game rules with respect to any kind of external world is exemplified by the game of chess. The identity of the various pieces, such as kings and queens, is defined by the strategic movements they are allowed to perform on the board, and these movements have nothing to do with the properties of their real-world namesakes: the king could be renamed "trout" and the queen "carrot" without affecting the game.

This nonmimetic character of the rules of games can be applied to the case of the literary text on two distinct levels. On the first, literature is a game because, like every specialized genre of discourse, it is governed by its own arbitrary (read: "could be different") conventions, which form what Juri Lotman calls a "secondary modeling system."[5] Of all these second-order literary rules, none is more basic than the convention that allows the language of fiction to create its own world without being held accountable for the truthfulness of its declarations with respect to reality. On the most elementary level, then, it is the declarative power of fictional language that supports the text-game analogy. To express this power, Iser proposes to replace the notion of literary representation with the concept of play. The concept "has two heuristic advantages: (1) play does not have to concern itself with what it might stand for, and (2) play does not have to picture anything outside itself" ("Play of the Text," 250).

On the second level—if we accept the poststructuralist interpretation of Saussure—the nonmimetic character of literature is not a choice but the unavoidable consequence of the medium's arbitrary nature. As Philip Lewis argues, "Playing with words may . . . involve more than juggling them about, re-orienting their interrelationships; it may compromise their very status as linguistic signs, revealing their inadequacy to relay accurately or completely the 'realities' they purport to designate" (quoted in Motte, *Playtexts*, 21). In this perspective, literary wordplay is not a specialized use of language contrasting with serious utterances but an allegory of the very condition of signification.

During the past twenty years, as structuralism gave way to deconstruction, the focus of interest in ludic metaphors has gradually switched from *ludus* to *paidia*, which means it has shifted from considering the notion of game as rule-governed activity to seeing play as subversion of rules. For A. J. Greimas, the quintessential structuralist, literary signification manifests itself in spatial configurations of semes, such as oppositional axes, triangles, and especially squares, and the rules of the textual game are revealed in the visual mapping of its semiotic structures. The narrative development of the text brings about permutations of elements, but to remain intelligible, the global pattern must retain at all times an axis or center of symmetry. In a seminal essay, "Structure, Sign, and Play in the Discourse of the Human Sciences," Derrida describes as follows the configuration of the structuralist structure.

> The function of this center [of a structure] was not only to orient, balance, and organize the structure—one cannot in fact conceive of an unorganized structure—but above all to make sure that the organizing principle of the structure would limit what we call the freeplay of the structure. No doubt that by orienting and organizing the coherence of the system, the center of a structure permits the freeplay of its elements inside the total form. And even today the notion of a structure lacking any center represents the unthinkable itself. Nevertheless, the center also closes off the freeplay it opens up and makes possible. *Qua* center, it is the point at which the substitution of contents, element, or terms is no longer possible. At the center, the permutation or the transformation of elements (which may of course be structures enclosed within a structure) is forbidden. At least this permutation has always remained *interdicted.* (247–48)

To the self-contained structure of games, in which movement is limited to specific slots determined by rigid rules, Derrida opposes a "free play" *(jeu libre)* of elements in a decentered, self-transforming, fluid organization. One of the meanings of *jeu* in French (also found in the English *play*) is the space between two pieces that occurs when one of them is not properly tightened up. This space, which allows movement, can be taken to represent the lack of fit between language and the world and the instability of linguistic signs. Free play arises because there are no "transcendental signifieds," no rules that fix meaning by anchoring language in an extralinguistic reality. Another conceivable interpretation of *jeu* is the uncertainty of the boundaries that circumscribe the semantic territory—or value—of linguistic expressions. In a purely differential conception of language, the value of word *A* ends where the value of *B* begins, but since these values are not backed by positive referential relations to a realm of extralinguistic objects, the exact location of the line of demarcation is undefinable.

It is always hazardous to try to paraphrase Derridean terminology, especially with a concept as broad as free play, but without taking an excessive interpretive risk, we can say that it involves the following ideas:

- The rejection of stable meanings in favor of an emergent conception of signification.
- The impossibility of achieving a totalizing and definitive apprehension of the literary text.
- The rejection of binary oppositions between "serious" and "playful," factual and fictional uses of language. Since *jeu* is inherent to the structure of language, its emphasis in literature is an allegory of the fundamental conditions of signification.
- A conception of playing that gives initiative to the multiple meanings of words and to their ability to activate chains of reactions, so that playing, for both author and reader, becomes indistinguishable from "being played" by language.

Another leading figure of postmodern thought who associates writing—all kinds, but especially literary—with the concept of *jeu* is Michel Foucault. In "What Is an Author?" he writes, "Referring only to itself, but without being restricted to the confines of its interiority, writing is identified with its own unfolded exteriority. This means that it is an interplay of signs arranged less according to its signified content than according to the very nature of the signifier. Writing unfolds like a game *(jeu)* that invariably goes beyond its own rules and transgresses its limits" (120).

In this passage the playful nature of writing is associated with three features:

- Lack of reference to an external reality, which leads to self-reference (since language must refer to something).
- Priority of the signifier over the signified as a principle of organization (a claim that privileges phenomena such as puns, rhymes, and alliteration at the expense of metaphor, narrative structure, and the recurrence of motifs and themes).
- The subversion of rules.

While a game necessarily involves players, the interpretations of the game metaphor surveyed above differ widely in their implicit view of the player's identity. In the structuralist conception, the forces at play in the organization of the text are the "elementary structures of signification," which can be seen either as the structure of language itself or, in a Chomskyan framework, as the hardware of the brain. In Oulipian games, as well as in the fixed forms of folklore, the player is the author, the plaything is language, and the reader's mode of involvement is mainly that of a spectator or referee. Derrida's notion of free play emphasizes the agency of language, and it regards users as mediators of play rather than as plenary participants. Through their attention to the horizontal/differential relations that make up the texture of the text and the system of language, author and reader release semantic energies that are "always already" contained as virtualities in the substance of the medium. It is only when the reader is cast in the player role—as he will be literally in computer games, figuratively in texts with variable sequentiality, and programmatically in theories of the reader-as-author—that the game metaphor supports a poetics of interactivity.

Comparing the Game and the World Metaphor

With these general observations in place, we can proceed to the systematic comparison of the game and world metaphors. The entries below, as well as table 2, summarize the main points of the preceding discussion in a contrastive manner.

Function of Language

The text-game analogy borrows features from many games, but if one game stands out as a prototype for postmodern aesthetics, it is the game of construction (a game that, paradoxically, eludes classification in Caillois's categories, unless it is regarded as mimetic, a view that does not sit well with postmodern aesthetics). According to Roland Barthes, reading is a "cubist" exercise in which "the mean-

Table 2. Feature comparison for the metaphors "the text as game" and "the text as world"

	Game	World
Function of language	Cube, matrix, toolbox	Mirror (virtual image), picture
Substance of language	Opaque, visible	Transparent
Meaning	Relational	Referential
	Horizontal	Vertical
	Fluid, emergent	Textual information needs to be supplemented by imported knowledge
	Entirely contained in text	
	("Il n'y a pas de hors-texte")	
Reader's attitude	Reflexive	Nonreflexive
	Lucid	Willing suspension of disbelief
	Refusing illusion	Accepting illusion
Type of activity	Surfing the surface	Exploration
	Construction, permutation, transformation	Voyeurism
Form	Form as exoskeleton	Organic unity of form and content
	Emphasis on arbitrary formal constraints	
Role of chance	Ambiguous:	Negative (words express how things are in fictional world)
	Positive (words take initiative)	
	Negative (signs must be overdetermined)	
Conception of space	Space occupied by text: Figures, arrangement on page, network of accessibility relations between units	Space represented by text: Environment, landscape, geography
Requirements	Specialized "literary" competence	General linguistic and cultural competence
		Basic life experience
Critical analogy (Barthes)	"Writerly"	"Readerly"

ings are cubes, piled up, altered, juxtaposed, yet feeding on each other" (*S/Z*, 55). Milorad Pavić develops the constructivist metaphor in the introduction to his *Dictionary of the Khazars*, a novel presented in dictionary form: "Each reader will put together the book for himself, as in a game of dominoes, or cards" (13). In this perspective, the text is "open" and reconfigurable, a matrix containing potentially many texts, a network of relations between semiautonomous units; it is a toolbox rather than an image, a renewable resource rather than a consumable good.

In the regime of the text as world, by contrast, the role of language is best compared to a mirror. (See the epigraph of chapter 13 of *Le Rouge et le noir* by Stendhal: "Le roman: c'est un miroir qu'on promène le long d'un chemin" [A novel is a mirror that one carries along a road].) Rather than focusing on the flat surface of the mirror, the spectator looks into its depth, where she discovers a three-dimensional reality. The metaphor of the text as world is linked to a phenomenology of "as if," of passing as, of illusion.

Substance of Language

If letters and words are the pieces of a construction game, they are opaque objects, and attention is directed toward the visible or phonic aspects of language, the material substance of the medium. Rhymes, alliteration, acrostics, palindromes, and calligraphic effects are among the language uses that are most commonly described as play. On the other hand, if the text is a mirror, words are transparent signs. Their function is to be a passport to the fictional world, to transport the reader into an alternate reality. Once the fictional world becomes present to the imagination, the language that monitored the mental simulation is partly forgotten.

Meaning

The game metaphor supports a Saussurian conception of language in which signs acquire their meaning not from vertical relations with objects in the world but from horizontal relations with other signs. The literary text, like language itself, is a self-enclosed, self-regulating system in which meaning is determined by a strategic configuration of elements.[6] The conception of meaning associated with the game metaphor can also be described as the product of a field of energies. Meaning is not a preformed representation encoded in words and in need of decipherment but something that emerges out of the text in unpredictable patterns as the reader follows trails of associative connotations or attends to the resonance of words and images with the private contents of memory. This operation is like following links on the Internet: surfing the surface, remaining in perpetual motion.

In the world metaphor, meaning is vertical, since language refers to the objects of the fictional world. As Thomas Pavel has observed, the rise of the possible-worlds approach brought an end to the "structuralist moratorium on representational topics" (*Fictional Worlds,* 6), an important part of which is the concept of reference. As we saw in chapter 3, in a vertical conception of meaning the primary role of language is to direct attention toward objects in the textual world, to link them with properties, to animate characters and setting, and to lure the imagination into narrative simulation.

Reader's Attitude

Playing a game is a lucid activity; the player must see through the deceptive moves attempted by the opponent. The attitude favored by the text-as-game aesthetics is therefore one of critical distancc. The reader is not allowed to lose sight of the materiality of language and of the textual origin of the referents. One of the narrative strategies most widely described as play is the self-reflexive, anti-immersive metafictional stance through which authors such as Laurence Sterne, Denis Diderot, Flann O'Brien, or John Fowles remind readers that they are dealing with a constructed plot and not with life itself. This ideal of demystification stands in stark contrast to the suspension of disbelief (or pretended belief) that describes the attitude typical of the world metaphor. Here the reader pretends that there is a reality existing independently of the language that creates it. The pleasure of the text depends on the reader's willingness to bracket out objective knowledge and surrender to illusion.

Type of Activity

What do you do when you play with blocks? You arrange them into various configurations. You permute, transform, try out all the combinations. You build and you create. Contemporary critical idiom calls this activity the construction of meaning. In a game known as deconstruction, the blocks can also be taken apart to see how meaning is put together.

What do you do when you visit a new world? You explore its territory, and you spy on its inhabitants. You act, in other words, like a tourist and a voyeur—two roles that are generally not associated with intellectual sophistication. This might in part explain the prejudice of many academic critics against immersive reading.

Form

The text as world upholds the classical ideal of an organic unity of form and content. The form is dictated by the content and should not attract attention to itself. It supports the text like an internal skeleton.

The text as game treats form as an exoskeleton: rather than subordinating form to content, it treats content as the filler of the form. As we have seen, games are constituted by arbitrary rules, and literary texts imitate games by subjecting themselves to pragmatically unnecessary formal constraints.

Role of Chance

The ambiguous role of chance in the text as game has already been discussed. In Surrealist aesthetics, the random encounter of words creates poetics meaning (cf. Lautréamont's conception of beauty as "the chance meeting on a dissection table of a sewing machine and an umbrella," as well the predilection of the movements for mad-lib party games and their "exquisite cadavers"), while in Oulipo aesthetics the purpose of constraints is to purify the text from randomness. In the text as world, chance plays a clearly negative role, since the function of language, for those who accept the illusion, is to provide a faithful image of an independently existing reality. While the author, as world-creator, is free to invent, the narrator is bound by the facts invented by the author, and his choice of words must satisfy truth conditions with respect to the fictional world. If he fails these conditions, he is considered unreliable.

Conception of Space

In the text-as-world metaphor, space is a three-dimensional environment to be lived in, an area for travel, a landscape and a geography to be discovered in time. It is mapped by the bodily movements of characters from location to location. In the regime of the text as game, space is the two-dimensional or even three-dimensional playing field on which words are arranged (page or screen),[7] and its map is the network of relations (analogies, oppositions, electronic links) that connects textual units, determines patterns of accessibility, and traces formal figures.

Requirements

In order to play a game, the player must be thoroughly familiar with the rules. The text-as-game metaphor is esoteric and elitist; readers need literary competence to appreciate the text. According to Jonathan Culler (*Structuralist Poetics*), this competence is so specialized that it must be learned like a second language, preferably in school. The text as world is a much more populist conception. All the reader needs to gain access to the fictional world is a basic knowledge of language, life experience, and reasonable cultural competence. If there are rules to learn in order to navigate the textual world, these rules can be learned on the fly.

Critical Analogy

I borrow from Barthes the concepts of writerly and readerly. It is no secret that Barthes regards the readerly as an inferior category; in his view, the sophisticated reader needs to graduate from the readerly to the writerly to attain higher forms of pleasure. The readerly frames the reader as a passive consumer who devours the text and throws it away: "This reader is . . . plunged into a kind of idleness—he is intransitive; he is, in short, serious: instead of [playing] himself,[8] instead of gaining access to the magic of the signifier, to the pleasure of writing, he is left with no more than the poor freedom either to accept or reject the text" (*S/Z*, 4). The writerly, by contrast, is seen as promoting an active and playful participation of the reader in the act of writing: "Why is the writerly our value? Because the goal of literary work (of literature as work) is to make the reader no longer a consumer, but a producer of the text" (ibid.). In a later work, *Le Plaisir du texte*, Barthes contrasts *jouissance* (bliss), the experience provided by the writerly, with *plaisir* (pleasure), the experience of the readerly, and though he considers the two experiences to be qualitatively different and therefore complementary, he also insinuates, in a prescriptive spirit characteristic of the proponents of literary avant-gardes, that *jouissance* is an "extreme," unspeakable (*indicible*) form of pleasure that is diffused throughout the text, while *plaisir*, a "little jouissance," depends on anticipation and occurs only at climactic moments (34–35).

Can the game and the world aesthetics be reconciled? The polarity of the two columns of table 2 suggests a fundamental incompatibility of the two metaphors. A world is not a game, but as postmodern literature has ingenuously demonstrated, textual worlds can be turned into tokens of play. Brian McHale (*Postmodernist Fiction*, 6–11) defines postmodernism as a movement that foregrounds ontological preoccupations, as opposed to modernism, which was dominated by epistemological questions such as "Who am I?" "Can I know the world?" and "Can I know myself?" Postmodern fiction does not directly ask, "What is the nature of being?" but it thematizes ontological problems by treating worlds as toys, by juggling them in the textual space, by building alternative ontologies, by playing with transworld identity, by transgressing ontological boundaries, by making worlds morph into other worlds, and by merging generic landscapes.

The consequence of this play is that immersion becomes thematized. But since immersion is a state of forgetting language and losing oneself in the textual world, its thematization also means breaking the spell that makes it possible. It takes deprivation for the reader to come to realize the importance of immersion. The many-worlds texts of postmodernism offer glimpses of what it means to settle down in a world, but as soon as the reader develops a sense of belonging, these texts break the illusion or transport her to another world. By shuttling the reader back and forth between worlds, by constantly shifting perspectives, by proposing multiple realities that relativize each other, by constructing and voiding worlds—as it exposes their language-made nature—the postmodern text keeps the reader in a state of permanent jet lag.

Fortunately for those who prefer to grow roots in a textual world, the frenetic world-play of postmodernism is not the only way to reconcile the two metaphors. The best compromise of all is simply to regard the concepts of game and world as complementary points of view on the same object, much in the way modern physics uses the metaphors of wave and particle as alternative conceptualizations of light. This is not to deny that some texts are inherently more gamelike (hypertext, visual poetry, postmodern novels) and others more worldlike (realistic texts). But while the nature of the text usually favors one of the conceptualizations, each of the two metaphors provides a point of view from which we can observe features that remain invisible from the other. If we are unable or unwilling to switch perspective, we will never appreciate the language games that are being played in textual worlds or the worlds that are being manipulated through language games. Yet because an observer cannot simultaneously occupy two different points in space, the complementarity of the two metaphors also means that we cannot experience both dimensions at the same time. We must therefore immerse and deimmerse ourselves periodically in order to fulfill, and fully appreciate, our dual role as members of the textual world and players of the textual game.

SIX

Texts without Worlds: Dysfunctionality as a Form of Play

In the sense that I adopt in this book, immersion is the response to a text, whatever its medium, that is able to conjure the presence of a world to the imagination. This conception of immersion is honorific, since some representational texts are unable to bring their world to life, but it does not necessarily regard immersion as an exclusive source of artistic merit. A text can indeed achieve aesthetic status by spurning world-creation and by deliberately preventing immersion. In this chapter, I will look at digital texts that actively pursue this agenda, sacrificing the "world" aesthetics to the "game" aesthetics.

To play with an object—the way, for instance, a child might drum on a cooking pot or use a manhole cover as the house of the wolf—deprives the object of its usual function and creates a new use. Huizinga's definition of play (discussed in chapter 5) stresses its lack of practical usefulness: play is an activity connected with no material interest, and no profit can be gained from it. Play is therefore a dysfunctional activity by comparison with the routine actions of everyday life that guarantee our material survival, but it is highly functional within its own "magic circle," as Huizinga called the space set apart by the game's rules.[1] Just as objects become toys by losing their real-world function, texts achieve playfulness through dysfunctionality with respect to practical reality. The reliance of digital texts on the computer, a highly functional tool, takes dysfunctionality to unprecedented levels and opens a wide range of possibilities, since computers have many practical functions, each of which can be subverted in a distinct way. But the love affair of digital art with dysfunctionality does not mean that it is dysfunctional as art, nor that it entirely lacks functionality; on the contrary, art that ironically flouts one of the uses of the computer can lead the audience not only to an appreciation of the cleverness of the author, an eminently aesthetic experience, but also to a critical awareness of the mode of operation of digital technology or of its role in society.

Dysfunctionality is admittedly a relative feature. If one assumes that the most practical function of language is to communicate efficiently information about the real world, then it follows that narrative fiction is dysfunctional by comparison with news or historical narrative, hypertext fiction is dysfunctional with respect to linear storytelling, and concrete poetry is dysfunctional compared to the lyrical kind. The texts I discuss in this chapter are all dysfunctional with respect to standard narrativity and fictional world-creation; in addition, some of them satirize particular applications of digital technology (video games, tools) and its mode of operation (code, interfaces), and they may go as far as rendering language itself dysfunctional.

Dysfunctional Code

Of all the levels of computer operation that inspire playful activity, code is the most elementary. Ideally, code should perform a useful task in a reasonably economic way (i.e., a way that does not take too much processor time or memory), and it should be written in a way that other people can easily understand, so that future programmers can easily modify it. The transgression of these criteria is the aim of "codework," the use of computer languages for aesthetic purposes.

There are many different types of codework, and they present different forms of dysfunctionality. All standard programming languages presuppose a universal machine, that is, a machine able to perform every task that is computable. This reliance on a common standard makes it possible to perform the exact same task (for instance, printing "hello, world") through different languages or through different programs written in the same language. Computer languages, in contrast to the natural kind, are therefore perfectly translatable into each other. This translatability has given rise to two forms of play with code.

The first of these two types consists of writing needlessly complicated programs in a standard language, such as PERL, Java, or HTML. The resulting obfuscation is not inherent to the language but rather derives from a particular use of this language; Saussure would assign it to *parole* rather than to *langue*. Here is how Nick Montfort, who has thoroughly studied the issue, describes the aesthetics of obfuscation:

> All obfuscation—including naming obfuscation [i.e., the choosing of the names of variables and constants by the programmer] as well as language-specific ones, such as choosing the least well-known language construct to accomplish something—explore the play in programming, the free space that is available to pro-

> grammers. If something can be done one way, it cannot be obfuscated. It is this play that can be exploited to make the program signify on different levels in unusual ways. (Montfort, "Obfuscated Code," 197)

Another form of play with code is the invention of "weird languages," that is, of languages that can do everything a well-designed computer language does but in a way that is almost incomprehensible to the human programmer. While efficient computer languages use key words whose meaning is transparent because they read like English words (e.g., repeat, go to, begin, end, if . . . then), obfuscated languages may double-code meanings, so that, to take an imaginary example, *purple* could stand for *subtract* or *kiss good-bye* for *jump* to a certain address encoded in *good-bye*. An example of this practice is the computer language Shakespeare, which is read differently by humans and by the computer. (Here I paraphrase the description of Mateas, "Weird Languages.") For human readers, words maintain their normal English meaning; but for the computer, every word corresponds to a specific value or executable operation. The names of characters from Shakespeare's plays are interpreted by the computer as variable names; all positive nouns (*lord, angel, joy*) are constants with the value +1; negative nouns (*bastard, beggar, codpiece*) have the value –1, and adjectives before nouns multiply their value by 2. The scene indications are locations for jumps. Verbs stand for operations; for instance, enter and exit statements indicate which variables are active at a given point, and *return* performs a go-to statement. It takes a list of translations to turn the dialogue into executable code since the compiler could not decide on its own what counts as a positive and as a negative noun. The purpose of the exercise is to create a program that runs without crashing, that performs some kind of task (asking for a useful task would set the bar too high), and that resembles a play on the surface level. In his discussion of weird languages, Michael Mateas quotes a fragment of a Shakespeare program that "reads an input and prints it out in reverse order" ("Weird Languages," 270):

> *[Enter Othello and Lady Macbeth]*
> OTHELLO: You are nothing!
> Scene II: Pushing to the very end
> LADY MACBETH: Open your mind! Remember yourself.
> OTHELLO: You are as hard as the sum of yourself and a stone wall. Am I as horrid as a flirt-gill?
> LADY MACBETH: If not, let us return to scene II. Recall your imminent death!
> OTHELLO: You are as small as the difference between yourself and a hair!

While Shakespeare remains operational, Mezangelle, the pseudo computer language invented by the Australian digital artist Mez is a purely visual attempt to imitate computer code. Here is a sample of Mezangelle:

```
15($stiff . ening with fea((c))r(eam)!)
5($limb . less f((l))ailings)
53($swelt . (i)ered.fractures
35($uglin((dr))ess(es) (x)
30($denied . (see)d(s)ire
331($snipped-genital-(pup(a))pets x))))))
object
33($rot-cavity-m(f)o((a)rm)ldings)
3($(flesh . g)r(e)ying . orGa(mete)n(Elles))
13($let ((form))
13($get* bindings . b(lind)ody))[2]
```

The Shakespeare program tries to hide that it is a piece of computer code by imitating a dramatic script. Here it is the other way around. The poem tries to pass as computer code by imitating the syntax of a computer language (it most resembles LISP) and by using special graphic elements, but it cannot be executed: there is no compiler for this pseudo language. The purpose here is not to communicate with the machine but rather, following a tradition that runs from nineteenth-century Symbolism to twentieth century Dadaism, Surrealism, Lettrism, and concrete poetry, to produce a new poetic idiom. The use of parentheses is symptomatic of an ambition to fight the linearity of language by rendering symbols semantically polyvalent. For instance, in the above example, "see" can also be read as "seed" or "seeds," and "seeds" overlaps with "(d(s)ire." Whether or not these language experiments produce dysfunctionality or a functionality of a higher order depends on how much effort readers are willing to devote to their decoding and on whether this effort is found worthwhile. Similarly, there are people who find the language of *Finnegans Wake* highly dysfunctional, while others admire it as a synthesis of different languages that overcomes the disaster of Babel. (Incidentally, Mezangelle has been widely compared to *Finnegans Wake.*)

Dysfunctional Tools

Another form of playful dysfunctionality affects the level of computer programs. Here I discuss two examples. In the first of them, Adrian Ward follows in the footsteps of Swiss sculptor Jean Tinguely, who created mechanical contraptions that crank their wheels without doing anything useful, by subverting the utilitarian spirit of commercial drawing software. The graphic tools of the program are turned into autonomous agents with a will of their own. We take it for granted that in a graphic program the code listens to our input; if we select the straight-line tool, we do not expect the program to draw an arabesque. *Auto-illustrator* breaks this basic contract between the software designer and the user by complicating (rather than completely severing) the relationship between the movements of the hand and the behavior of the tools. The program does not listen directly to the position of the cursor but translates it into output through a mysterious algorithm. If you select the freehand pencil tool, the system draws unpredictable (though rule-governed) graffiti; if you select the text tool, the system picks the letters and invents nonsense words. The square and the oval tools do not draw regular geometric shapes but give you a choice between "shabby" and "precise" shapes, as well as between "childish," "artistic," and "regular." (The illustration of figure 2 shows a combination of childish and shabby.) There is also a bug tool, which places moving creatures randomly on your screen. They will create art for you by crawling around and drawing lines. If you don't like the result, there is an exterminator tool that will let you get rid of the creatures.

Auto-illustrator's play with code does not take the form of making it directly visible but rather of asserting the artistic dimension of the programmer's activity. In other words, it is not codework but what Christiane Paul calls "software art" (*Digital Art,* 124). In an article included in a user's guide to *Auto-illustrator* that is no longer available, Florian Kramer observes that in commercial applications, "programmers are frequently considered to be mere factota, coding slaves who execute other artist's concepts." Software art liberates programmers from the tyranny of corporate work by letting them express their own vision, using code as a metamedium to control other media: language, sound, color, shapes, and animation.

Richard Powers's Web-based story "They Come in a Steady Stream Now" (no longer available) constitutes an exception with respect to the other texts discussed in this chapter in that it builds a semblance of world and retains some degree of narrativity, but it shares with them a playful attitude toward computer technology. The work satirizes the proliferation of spam in e-mail, arguably the most widely used computer application, through an interface that simulates the arrival of messages in the user's mailbox (figure 3). Of the seventeen messages that reach us during the run of the program, ten are spam and seven "legitimate." The junk mail runs the familiar gamut of pornography, drug offers, and investment opportunities. In addition to the junk mail, the mail program is plagued by pop-up ads,

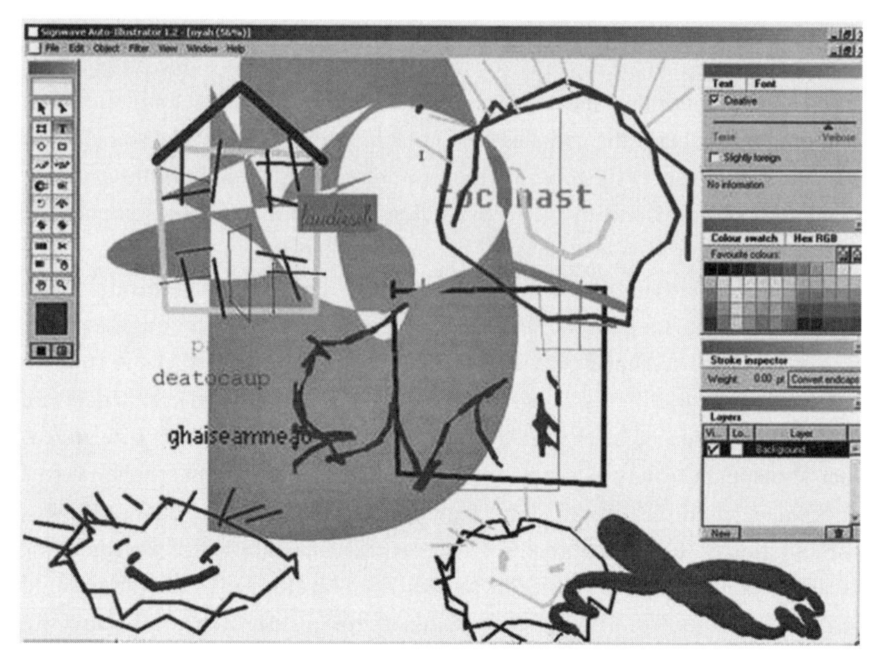

Figure 2. Auto-Illustrator, by Adrian Ward

which readers must close one by one before opening a new message. In contrast to the humor of the junk mail, the seven legitimate letters, addressed to the reader by a narrator named Richard Powers, contain a melancholic meditation on aging triggered by the junk mail's incessant hawking of drugs that promise to reverse the damage of time. In the last of the seven letters we read: "PLEASE REGISTER. The content you requested is available only to registered members. Registration is FREE and offers great benefits." The readers who dare to follow these instructions, mindless of the risk of viruses, by giving their e-mail address are rewarded with a message from Richard Powers in their own mailbox. In it they find a link to a PDF file that can be downloaded and printed. This file contains the text of the previous six simulated mails, together with a very Proustian conclusion in which the present absorbs the past and the past becomes present, allowing the narrator to relive an episode of his childhood. By including all the previously read installments, the final delivery contrasts the reading experiences of the digital and print versions. In the e-mail simulation, the text comes to the reader as a collection of fragments that create distraction through their many windows, through frequent interruptions, and through the obsessive need to click. The printable text gives rise to an entirely

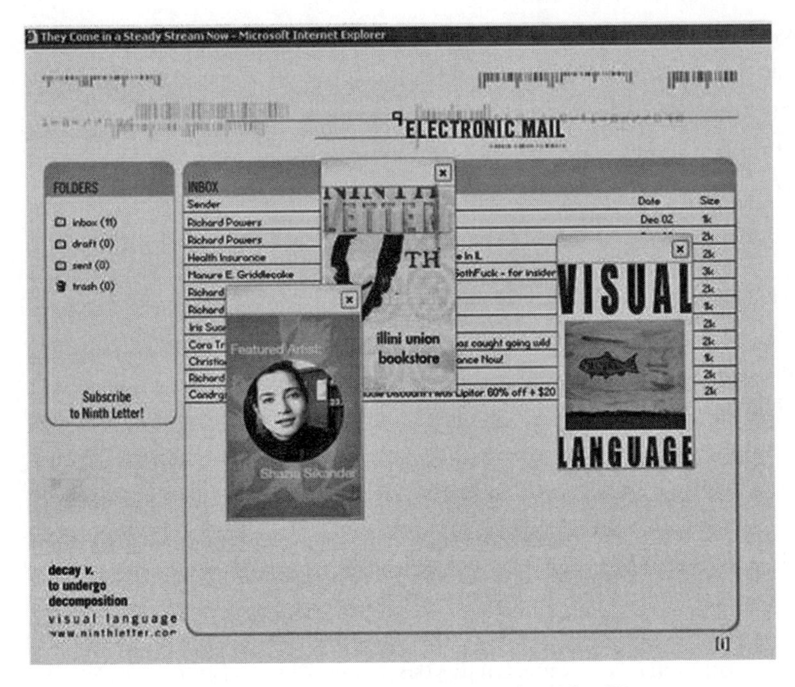

Figure 3. They Come in a Steady Stream Now, by Richard Powers

new reading experience. Now we can hold the entire text in our hands, enjoying a haptic relation with it which is absent from reading on a screen; we can read it without interruption, and we do not have to worry about competing windows. All these features, by freeing our attention from the interface and from the material conditions of reading, enable us to pay greater attention to the semantics of the text and to the poetic quality of language. (It also helps, of course, that we are rereading rather than reading for the first time.) The originality of Powers's achievement lies in the complementarity of the comic experience of the screen version and of the lyrical experience of the print version. In its play with two media, the text manages to combine the power of print literature to induce emotions with the cleverness of conception and innovative presentation that we have come to expect of digital art.

Dysfunctional Games

Digital artists frequently borrow the mechanisms of video games to satirize their violent themes, sexual stereotyping, and above all their blatant commercialism. For instance, the project "Velvet Strike" (Schleiner et al.) uses the engine of the shooter game "Counterstrike" to spray antiwar slogans on the walls of the game environment (Paul, *Digital Art,* 203). A more literary work, "The Princess Murderer" (geniwate / Deena Larsen) fuses the common video game tasks of rescuing princesses and killing enemies by having the user identify with Bluebeard, the notorious wife-killer. Simulating a common video game feature, the interface displays a bar that represents the number of princesses remaining to be killed, but the number varies randomly, rather than reflecting the player's achievements (Ensslin, *Literary Gaming,* 92–101). More game than "The Princess Murderer," Jason Nelson's "Game, Game, Game and again Game," borrows a gameplay made familiar by the Super Mario Brothers series to create a "game/poetry/art monstrosity,"[3] to which I devote the remainder of this section (figure 4).

The player (represented by a buglike blob) moves horizontally on the screen by hitting the left or right arrow keys. When there is no platform to support the player, she can jump by hitting the space key. If she falls without reaching another platform, she dies, and a solemn voices announces: "You are going to meet your maker." The goal is to reach a door that takes the player to the next level. At the beginning of each level, the screen contains only a few objects, but as the player moves around, she unlocks texts, images, and animations that create a rich sensory and informational landscape. On this basic ludic scheme, Nelson piles up multiple forms of dysfunctionality.

- Difficulty is inconsistent: some levels are so easy to pass that the player has no time to read the written messages nor to appreciate the graphics. Others infuriate the player with their difficulty: I personally gave up on level 9 (out of 13) after meeting my maker one too many times.
- Gameplay is frustrating: there is no "save game" feature, nor any mechanism that gives players random access to the various levels. In order to make new attempts to get past level 9, I had to start each time from the beginning and solve once again the mazes of the previous levels. On a given level, every time the player meets her maker, she must start again from the same position.
- Graphics are unprofessional: the game subverts the smooth realism of the computer-generated interface of commercial video games with clumsy (but charming) childlike drawings that bear the imprint of the hand that drew them.
- Timing is defective: each level comes with a theme that addresses one of the pet peeves of the author: themes like "faith," "real estate," "capitalism,"

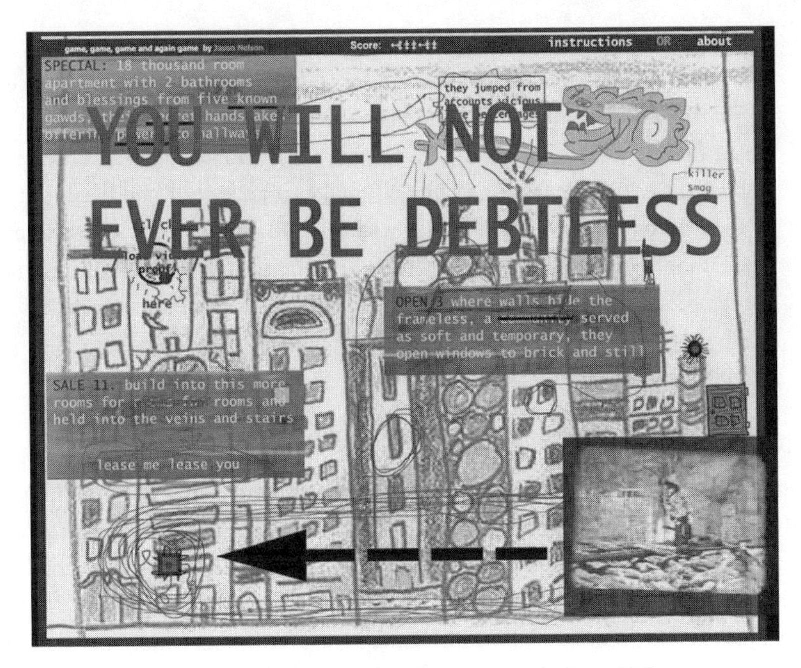

Figure 4. Game, Game, Game and Again Game, by Jason Nelson

"life coach," and "tourism." But these themes are sometimes flashed so quickly that the player has no time to read them.

- Score keeping is nonsensical: rather than giving the player useful information, it consists of meaningless strings of arrows pointing in arbitrary directions.
- The sound track is irritating: in addition to hearing the same message about meeting her maker over and over again, the player must put up with obsessively repetitive syncopations of digital sounds.
- The rewards are phony: just before the player reaches the door that will free her from the screen, she can click on a button that launches the kind of home video that was all the rage in the eighties: a grainy image of suburban family life that inspires nostalgia for obsolete technologies (or maybe relief over their passing).
- The text is absurd, and gameplay gets in the way of reading: as the player moves across the screen she unlocks bits of "poetry" that look more like the products of an automated algorithm than like the creations of an intelligent mind. If she takes the time to read (temporarily forgetting the

> game goals), she will find amusing nonsense in the mad-lib tradition, but the text will frustrate any desire for thematic or narrative coherence. Yet these "mash-ups" or "assemblages" (as the work's presentation on the ELO [Electronic Literature Organziation] Web site describes them) bear the editing marks of a human hand, as if an author had tried to turn them into a coherent text. Here is a sample: "Before [scribbled over] her death, before her ~~tenth year~~ of life, before she completed her collection of soft versions of hard objects, worms for [illegible, maybe tires], socks for roads, she was hit by hovering angels [scribbled over, replaced by sighs] playing soccer in the sky."

With so many obvious defects, how could the game fail to entertain? Just as, in logic, two negations combine into a positive statement, the accumulation of dysfunctional features adds up to a uniquely comical experience, a pleasure made of curiosity, discovery, and surprise reminiscent of the excitement of a child (or adult) who opens the windows of an Advent calendar, discovering treasures that lie behind the surface, and filling the landscape with images that stimulate the imagination. As a digital artist, it is through humor that Nelson defends his creative ground against the commercial games produced by the massive resources of the entertainment industry.

Dysfunctional Interface

For all its frequency in the discourse of digital media, the term *interface* remains hard to define. For Janet Murray interface is "the perceptible parts of a computer program that stand between the code and the user" (*Inventing*, 426). For Carl Therrien, the term encompasses "the physical means to provide input in a system as well as the feedback produced by the system" ("Interface," 305). Here I use *interface* to designate the devices through which digital texts reveal themselves to the reader, such as graphic design, mode of interaction, and accessibility of the active elements that bring changes to the display. An efficient interface is easy to learn, and it allows the user to perform tasks with minimal effort, but when it becomes second nature and makes the computer invisible, user-friendliness clashes with the artistic ideal of defamiliarization. This is why a number of digital artists experiment with interfaces that prevent a fluid, unencumbered reading experience.

A digital author once told me that he expects a large proportion of the pages he writes never to be read. If we think of good reading as parsing a text with the eyes and the mind from beginning to end, pondering the contribution to a global meaning of every one of its words, then many digital texts transgress this criterion by inviting partial browsing rather than complete processing. *Chemical Landscapes: Digital Tales* by Edward Falco consists, for instance, of landscapes generated by manipulating chemicals in a darkroom and of texts inspired by these landscapes. But the texts dissolve so quickly on the screen—not unlike an image of an undeveloped film exposed to light—that the reader's eye can only parse a fraction of their words. It is possible to complete the reading by recalling the same screen, but by that time most readers have probably forgotten the meanings collected during their previous visit.

Another digital work that challenges memory is *The Jew's Daughter* by Judd Morrissey and Lori Talley. Morrissey describes it as follows: "*The Jew's Daughter* is an interactive, non-linear, multivalent narrative, a storyspace that is unstable but nonetheless remains organically intact, progressively weaving itself together by way of subtle transformations on a single virtual page." The editors of the ELO collection also insist on narrativity: "*The Jew's Daughter* is a work that renegotiates the concept of the hypertext to present a reconfigurative narrative." Both of these quotes presuppose a very loose conception of narrative, much looser than the one I adopt in this book. It is true that, at first sight, *The Jew's Daughter* reads like a narrative—more precisely, like a text extracted from a modern or postmodern novel. Consider the first page:

> Will she disappear? That day has passed like any other. I said to you, "Be careful. Today is a strange day" and that was the end of it. I had written impassioned letters that expressed the urgency of my situation. I wrote to you that that that [*sic*] it would not be forgivable, that it would be a violation of our exchange, in fact, a criminal negligence were I to fail to come through. To hand to you the consecrated sum of your gifts, the secret you imparted persistently and without knowledge, these expressions of your will that lured, and, in a cumulative fashion, became a message.

This passage presupposes an intricate set of relations between three characters—the "she" of the first sentence, the narratorial "I," and the "you" that functions as narratee. Were *The Jew's Daughter* a genuine novel, the text would soon associate the pronouns with individuated characters, explain the situation of the narrator, tell the reader why he or she has to write letters, what secret the narratee gave to the narrator, and so on. The text perversely multiplies questions, but because of the interface (and also because of the writing style), these questions will not be answered. When the reader mouses over a blue word on the screen, part of the text replaces itself, and part remains the same, but the replaced area is not marked, and the process is so fast that the reader does not know what is old and what is new. Rereading the whole page does not help because the new passage blends perfectly, visually and syntactically, with the rest of the text. It would take photographic memory to identify the changes. But even if the reader were able to compare the current and the previous page—an operation that the interface does not allow—it would be impossible to read the text as an account of events that take place in a temporal or causal sequence, since the two pages overlap in content. The events of page 2 (provided the text reports events) cannot be said to follow the events of page 1, and the text, consequently, cannot be read as a sustained story. As Lori Emerson observes, "the text is not particularly about anything" (*Reading Writing,* 159), even though it stretches over 608 pages. But the antinarrative interface hasn't prevented critics from finding all sorts of hidden meanings; of all the texts of the ELO's collection of digital literature, none has inspired more commentaries than *The Jew's Daughter.* For instance, Katherine Hayles sees it as an allegory (or simulation) of the neural mechanisms that underlie consciousness (*Electronic Literature,* 80), while Emerson reads it as a "comment on the bookbound page" and as symptomatic of "a poetics of failure" in electronic literature (156). (I prefer to call this poetics dysfunctional, because dysfunctionality is more compatible than failure with an intentional effect.) It does not matter, in most of these commentaries, what particular words fill the screen, since the interpretations focus on interface mechanisms and graphic appearance rather than on semantic substance. But I do not blame critics for lacking the patience to submit *The Jew's Daughter* to an extensive and close reading, since by using *The Jew's Daughter* as an example of dysfunctional interface, I am doing the same thing.

My last example of dysfunctional interface is *Grafik Dynamo,* a mock graphic novel by the Canadian Web artists Kate Armstrong and Michael Tippett (figure 5). True to its title, *Grafik Dynamo* is a constantly changing work. It loads narrative fragments written by one of the authors into speech bubbles or text frames and combines them in real time with images randomly captured from the Internet. The images came originally from a live feed to Livejournal, a social networking site where people keep illustrated blogs, but more recently they have been taken from Flickr, a site where people display their prize pictures. The work mimics the look of a comic strip by displaying three frames separated by what is known as "gutters" in the jargon of the trade. Every few seconds—barely long enough to read the text—a new image-text combination appears in one of the frames. Sometimes image and text come into view simultaneously and all at once; sometimes the text

Figure 5. Grafik Dynamo, by Kate Armstrong and Michael Tippett

appears first on a black background, and the image reveals itself slowly from top to bottom, one line of pixels at a time. The order of replacement does not follow a left to right sequence but jumps capriciously among the three frames so that the eye does not know where to look next.

Grafik Dynamo subverts the reading habits associated with graphic narratives in multiple ways. We expect the relation between image and text in every frame to be narratively significant, which means that it contributes to the construction of a coherent storyworld, but here the juxtaposition is the product of random selection, a procedure that cannot produce narrative meaning, except by extraordinary chance. A sequence of frames should represent the same world and correspond to a temporal sequence of events, but here there is no visual unity or semantic relation between the frames. The diegetic narration in the text boxes at the bottom of frames usually would provide the context for the dialogue or thoughts represented in the speech bubble, but here the narration and the dialogue are often disconnected—the content of the bubble may change, while the bottom text remains the same. Frames should be read from left to right, but here the eye is attracted to the frame affected by the most recent transformation, and this results in a parsing of the display that moves randomly in all directions. I could go on and on with these violations of familiar reading protocols. But what does *Grafik Dynamo* offer in exchange? Readers (or viewers) can concentrate on the text alone by reading it as a parody of "Superman" style comics, they can focus on the images by asking what kind of story people are trying to tell by posting them on the Web, or they can treat the random pairing of text and image as a stimulant for the imagination by trying to create their own stories out of individual frames. But the short-lived appearance of the data does not give the user sufficient time to construct a narrative context for every frame; reading in this case will not be a systematic processing of all the materials presented by the project but a highly selective activity. The best way to use *Grafik Dynamo* is to run it in a background window and to make quick grabs to look at its current state, hoping that randomness will occasionally produce interesting couplings.

Dysfunctional Language

In *Grafik Dynamo,* dysfunctionality resides in the algorithm that pairs language and image, but language itself remains spared. The most extreme form of dysfunctionality involves the disintegration of meaning, a disintegration that takes several forms. One of them is the simulation of cognitively impaired speech. The incoherence of the text of Stuart Moulthrop's *Reagan Library* hints at a brain suffering from Alzheimer's disease, while the progressive invasion of the input text of Noah Wardrip-Fruin's and Brion Moss's *The Impermanence Agent* by foreign elements randomly selected from the user's hard drive suggests the memory loss of Nana, the grandmother of the narrator. The prominence of impaired language in digital texts is easily explained by the relative ease of generating nonsense through erasures and aleatory procedures, compared to the difficulty of creating algorithms that produce logically well-formed discourse.

One step further in the semantic disintegration of language is the decomposition of signifying units into their nonsignifying minimal components. The founder of the Lettrist movement in poetry, Isidore Isou, claimed that the focus of poetry has shifted from the paragraph in Romanticism, to the word in Symbolism, and finally to the letter, starting with Mallarmé and continuing through Dadaism and Surrealism.[4] Lettrism, the culmination of this trend, is well represented in the ELO anthologies, especially through the work of John Cayley, its best-known practitioner. In many of his works (for instance, *Translation* in the ELO collection), Cayley experiments with an algorithm that morphs words into other words and languages into other languages by operating substitutions on the level of their individual letters. This mechanism limits reading to the occasional recognition of lexically well-formed combinations or to guessing what the next existing word will be, as letters fill in blanks in a process reminiscent of the game Wheel of Fortune. The reader observes a continuous oscillation between sense and nonsense, waiting, sometimes successfully, sometimes in vain, for the aleatory mechanisms of the generative code to create sentential meaning. However, the reader is unable to register all the intermediary stages between readable words, either because the letters roll too fast to be individually noticed or because it is impossible to pay close attention to more than one of the multiple replacement processes that take place simultaneously on the screen.

In Brian Kim Stefan's *The Dreamlife of Letters* (figure 6), similarly, but through different means, the user witnesses what Alan Liu (*Laws of Cool,* 8–9) would call the creative destruction of textual meaning. The author took an unconventional creative essay (itself very fragmented in its syntax) by the poet and feminist theorist Rachel Blau du Plessis, selected individual words, presented them in alphabetic order, and made these words dance on the screen in a noninteractive visual show that highlights their common leading letter. By his own admission, Stefan pursues a form of algorithmic art that is pure poetic event and in which narrative content does not distract from the scene of writing: "My sense is that algorithmic art has to become dramatic but without attendant fictional or 'mythic' narratives; algorithmic art has to provide an 'objective correlative,' to use T.S. Eliot's term, not in the form of stories but of *events* that recreate in the human observer the sensation of human bodily function such as sexual desire, physical tension, repulsion, psychic violence, etc." ("Against Desire," 12). But how can language induce such sensations without eliciting representations in the reader's mind?

As language dissolves into its elementary particles, it becomes a spectacle to watch instead of a text to read. Language turns into spectacle in the hallucinating urban landscape of animated signs of Tokyo or Las Vegas, especially for the visitor who cannot read the characters, in the strings of code that race down the screen in the opening scenes of the film *The Matrix,* and in the stunning visual patterns of symbols that continually undulate, ripple, explode, or implode into other patterns but offer no readable content in Giselle Beiguelman's *Code Movie 1* (figure 7), a work included, for some reason, in the Electronic Literature collection.

The spectacularization of the word is particularly frequent in digital installations, a form of art intended for relatively short visits. Since installations must maintain a steady throughput, for fear of creating a bottleneck in the exhibit, they cannot afford to let the visitor become immersed in reading. Many installation artists pursue the dream of a total language in which sound, shape, color, animation, spoken voice, and written text respond to the movements of the user's body. In this language, as Bill Seaman puts it, "the word is not valued in a hierarchy over other media elements or processes" (231)—it is just one signifying element among many others. But language is often the loser in this semiotic cornucopia. As Roberto Simanowski observes about *Text Rain,* an installation by Camille Ut-

Figure 6. The Dreamlife of Letters, by Brian Kim Stefans

terback and Romy Architruv in which visitors try to catch the letters of a poem that fall from the sky, "The letters have left language behind and turned into visual objects as part of a sculpture" ("Double Coding"). The user would have to read the poem in a separate document, through what Simanowski calls double coding, to appreciate why the artists chose *this* poem rather than another.[5] Making language literally visible has long been the ambition of experimental forms of verbal art, such as concrete poetry, calligrams, or, more recently, multimodal novels (cf. Mark Danielewski's *House of Leaves*), but these experiments pursue meaning through an interplay between the graphic appearance of words and their semantic value. With the full spectacularization of language, semantics no longer matters, words and letters become pure shapes, and the text takes the last step out of literature and into visual art. Language has become fully dysfunctional.

Why Dysfunctionality?

The fascination of digital artists with dysfunctionality can take many forms and be motivated by various goals. One of them is political. The social and economic systems of our time have become so dependent on digital technology that the sabotage of the computer systems on which they rely can be turned into a major form of political activism. Radical hacktivist groups such as the artist collective

Figure 7. Code Movie 1, by Giselle Beiguelman

Critical Arts Ensemble (CAE) advocates resistance to capitalist society by disabling the databases maintained by governmental and corporate forms of power. As we read in one CAE's manifestos: "A small but coordinated group of hackers could introduce electronic viruses, worms, and bombs into the databanks, programs, and networks of authority, possibly bringing the destructive form of inertia into the nomadic realm" (quoted from Liu, *Laws of Cool,* 364). The creation of efficient viruses, which requires the outsmarting of increasingly sophisticated protective software, constitutes a tour de force highly valued in the exploit-worshipping culture of computer hackers. Even though the purpose of viruses is to induce dysfunctionality, they partake of a beauty comparable to that of a chess move, a mathematical proof, or a clever invention.

Whereas political dysfunctionality asks, "How can I subvert digital technology to encourage critical thinking?," another kind of dysfunctionality that I call ludic grows out of the question "What can I do with this technology, other than what it was meant for?" It is the spirit of play that drives artists to create images out of the limited graphic capabilities of the ASCII code, to turn computer languages into poetry, to write short stories with Twitter messages, or to punch whole novels on smart phone keyboards. Common to all of these examples is the negation of real-world practicality and the creation of a new functionality—the autotelic and self-reflexive functionality of art. Ludic dysfunctionality is only dysfunctional insofar as it rejects the subordination of technology to material pursuits.

The close relation between the spirit of play and the spirit of art means that ludic dysfunctionality is difficult to separate from a more programmatic form that I call experimental dysfunctionality. According to Russian formalists, the role of true art is to create estrangement from our thinking habits. This is why major artistic innovations are usually greeted with scandal—a reaction that movements like Symbolism, Dadaism, Surrealism, and Situationism actively courted by devising ever new ways of "épater le bourgeois." Before a movement succeeds in imposing new forms, if it ever does, it is perceived as destroying existing forms. All experimental art, in this sense, is born dysfunctional, but in the case of electronic art, this may be more than a passing phenomenon. Alan Liu regards "being cool" as the leading aspiration of digital culture, and he views "creative destruction" (or destructive creativity) as the primary way to achieve this status. The main difference between ludic and experimental dysfunctionality is one of attitude and ambition, and these features are a matter of degree. Whereas ludic dysfunctionality is individualistic and humorous, experimental dysfunctionality is systematic and programmatic; whereas ludic dysfunctionality does not take itself seriously, experimental dysfunctionality wants to promote new forms of art and new ways of thinking: rhizomatic, networked, viral, and, paradoxically, nonbinary. Whereas ludic dysfunctionality lets its creations speak for themselves, experimental dysfunctionality tends to wrap its projects in elaborate theoretical statements that tell the user what they are supposed to mean.

With its contempt of the mythical "bourgeois" (aka the broad public), experimentalism easily turns into elitism, especially when a craft as esoteric as computer programming is concerned. In an age when even computer science majors admit to not knowing how computers work (Chandra, *Geek Sublime*, 37), the cult of the dysfunctional in electronic literature can be seen as rejection of the aesthetics of the user-friendly that dominates the software industry and that, according to its critics, hides the functioning of the machine behind what Emerson calls its "glossy interface" (*Reading Writing*, xi). By restoring visibility to the computer, dysfunctionality represents the revenge of the hackers, who previously owned cyberspace, over the general public who now crowds their territory and takes the computer for granted. Given the complexity of the operations that mediate between zeros and ones and the display on the screen, however, one may question whether dysfunctionality provides users a glimpse into the inner working of the machine or simply frustrates them.

A last type of dysfunctionality that deserves mentioning is due to the recalcitrance of code to do what the coder intended. Most readers of digital works have experienced situations where the text does not seem to work properly. This experience can lead to a feeling of distress, especially when users cannot tell whether the apparent problem is a feature, a bug, or simply a result of their inability to operate the textual machine. When the failure is due to the code and not the user, dysfunctionality is inadvertent. As an example of this situation, consider the interactive drama Façade by Michael Mateas and Andrew Stern, in which the player interacts through language with computer-created characters driven by an AI engine. Façade is about a couple of yuppies, Grace and Trip, who are successful in their careers but who come to realize during the play the disastrous state of their marriage. I am sure that the authors hoped to create a coherent dialogue system, and quite often the AI does the job, but in the following example, Grace and Trip simply ignore the words of their guest, Heather, who is played by the user.

HEATHER: Let's not talk about marriage.
GRACE: Heather, when you get married you are supposed to forgive and not think about the past, yes or no?
HEATHER: It's not your fault, Grace.
GRACE: Heather, you blame me for all of this.
GRACE: I need your opinion about my decorating.
HEATHER: You don't have any decorating talent.
TRIP: It always amazes me when Grace comes home after a day at work and starts decorating.
GRACE: It's just the artist in me trying to get out.

In the first example, Grace responds to the word "marriage" rather than to the whole of Heather's utterance; in the second she grabs "fault" and wrongly feels accused; in the third, neither she nor Trip respond to Heather's attempt to insult her—a reaction rather out of character with Grace's prickly personality. If the work as a whole maintains narrative sense despite frequently dysfunctional dialogue, it is because the human interactor plays the role of a relatively passive observer. The plot is driven forward by the canned responses of the system-created characters, much more than by the contributions of the player.

Another source of inadvertent dysfunctionality stems from the fact that many layers of software and many types of hardware are interposed between a project and its execution. A digital work is supposed to run on many platforms, and each of these platforms has unpredictable idiosyncrasies, resulting in slightly or widely different behaviors. (Computer games simplify the problem by being designed for a specific platform, such as the Sony Playstation, the Microsoft Xbox, and the Nintendo Wii.) Variations from system to system may concern the fonts of the text, the color of the display, how much of the screen the work occupies, the speed of animation, or the responsiveness of the system to user input. In the best of cases, the differences are unnoticeable or insignificant; in the worst, they destroy the artistic effect. The more a work relies on a precise coordination of kinetics, sound, graphics, and linguistic meaning, the more susceptible it is to unpredictable behavior. In the worst situations, an application may refuse to run under a certain operating system or on a certain type of machine and become totally obsolete. Dysfunctionality, like death for living creatures, will be the fate that awaits all digital works, until we solve the problem of preservation.

Why, in the end, is digital art so infatuated with dysfunctionality? It may have to do with the origin of new media in a highly functional technology—one that played a crucial role in the development of contemporary culture and economy—for if art is an aesthetic object and if aesthetics is a "purposefulness without purpose," as Kant defined it, then art cannot be subordinated to a practical end. To become an art machine, the computer must therefore be taken out of the world of business, of work, of science, and of everyday life.

But dysfunctionality in new media art is not limited to playing with inherently digital phenomena such as code, interface, and tools. In many of the examples discussed above, dysfunctionality affects language itself by making it nonsensical or illegible. How can one explain this sustained assault on the most fundamental, versatile, and powerful mode of signification—the one that makes us truly human? To qualify as *literary* art, a work must give a prominent role to language, even when language is combined with other types of signs such as sound and image; to qualify as *digital* art, the work must do more than transpose a printable text to the screen, it must take advantage of the distinctive affordances of its medium. But the contributions of the computer to language pale by comparison with its contributions in the visual domain. Computer graphics, digital photography, and animations have not only taken realism to new heights, they have also made possible genuinely novel forms of graphic art, from the mapping of fractals to the creation of visually stunning virtual worlds that users can explore. The repertory of medium-specific operations that computers can perform on language is much more limited. The networking capabilities of digital media—e-mail, Twitter, blogs, and Web sites—have greatly facilitated textual communication, but most of these uses focus on social relations rather than on aesthetic experience. Automated text-generation is still in infancy, and as a source of verbal art it has yet to deliver more than amusing nonsense, which means, precisely, dysfunctional language. It is only by making language dance on the screen, by changing the shape of letters, by exploding words into their elementary particles, and by making them appear and disappear in response to the user's voluntary or involuntary actions, in short, by making language into a live presence, that the computer can truly innovate with respect to print. None of these operations encourage semantic decoding, nor do they facilitate storytelling or world-creation. The strength of new media does not reside in producing purely language-based texts but in creating multi-modal texts and worlds that provide a rich sensory and imaginative experience. If digital texts are to achieve literary status in spite of their proclivity for multi-modality, they must attract attention to language, and depriving language of its functionality is one of the best ways to make it visible. Dysfunctional language is language that speaks to the senses and refuses to disappear, language that opposes its opaque material presence to the mind's desire to traverse it in order to grab meaning. The dysfunctional use of language in electronic literature may thus be a way to make virtue out of necessity.

Reading Dysfunctionality

The issue of reading, especially of reading digital texts, has recently been at the center of a lively conversation.[6] It was long assumed that the "correct" or at the very least optimal way to read literary texts was close reading, a type of scrutiny traditionally taught in literature classes that involves "detailed and precise attention to rhetoric, style, language choice, and so forth through a word-by-word analysis of a text's linguistic techniques" (Hayles, *How We Think,* 58). Close reading can be thought of as effortful, as slow, as involving a cognitive mode that Hayles calls deep attention, that is, "concentrating on a single object for long periods (say, a novel by Dickens), ignoring outside stimuli while so engaged, preferring a single information stream, and having a high tolerance for long focus times" (Hayles, "Hyper and Deep Attention," 187). It is the mode of reading promoted by New Criticism, and it works best with short texts, especially lyric poetry, which was regarded by New Criticism as the essence of literariness. The purpose of teaching close reading, in this perspective, is to combat "bad" reading habits, such as the "reading for the plot" inspired by the page-turners of popular narrative.

Critics have complained about the scarcity of close reading of digital texts. In the 1990s, scholars were so caught up in the novelty of the medium and in its perceived fulfillment of postmodern literary theory (see chapter 8), that they glossed over texts, praising the liberating potential of interactivity but paying scant attention to what the texts were actually saying. Or rather, the texts were invariably reflecting on their own form of textuality. To paraphrase Marshall McLuhan, the medium was the all too predictable message. In *Close Reading New Media* (2003), Jan Baetens and Jan van Looy made a plea for taking more seriously the individual meaning of digital texts; so did Roberto Simanowski in *Digital Art and Meaning* (2011). This plea has been answered by critics such as Astrid Ensslin, Alice Bell, and David Ciccoricco, who have applied narratological concepts to the analysis of such genres as hypertext, video games, and game-literature hybrids.

But close reading is not the only way to read. To the deep attention required by close reading, which she sees as typical of the print medium, Hayles opposes another cognitive style that she calls hyper attention and regards as typical of a generation that has grown up with digital media: "Hyper attention is characterized by switching focus rapidly among different tasks, preferring multiple information streams, seeking a high level of stimulation, and having a low tolerance for boredom" ("Hyper and Deep," 187). As a mode of reading, hyper attention has both positive and negative features: on the positive side, it allows readers to keep track of, and hopefully synthetize, various types of information, often presented in different windows; on the negative side, it smacks of attention deficit disorder (ADD). Hyper attention is the ability to quickly locate information on the space of a screen, an ability indispensable to video game players, but it is also the habit of doing homework while texting to friends and watching TV.[7]

Close reading comes up largely empty in the case of texts that deliberately implement one of the forms of dysfunctionality. One can describe how these texts function, as I have done in this chapter, but a description is no more a reading than a paraphrase or summary if by reading one understands something like interpretation. To close read a disappearing text, such as Edward Falco's *Chemical Landscapes*, or a dynamically self-modifying text, such as *The Jew's Daughter* or *Translation*, one would have to take screen shots, which would immobilize the page. Moreover, the method of close reading does not work with long texts, unless it cuts them arbitrarily into small sections. Who would want to pay detailed attention to every one of the 608 pages of *The Jew's Daughter*? That is certainly not what "regular" readers do with such texts. So what do they do? The kind of attention inspired by the dysfunctional texts I have presented in this chapter is better described as hyper than as deep because dysfunctional texts are a form of conceptual art. The trademark of conceptual art is that its formula must be entirely original. The generative idea resembles the punch line of a joke, in that viewers may or may not get it, but when they do, they get all of it. Conceptual art specializes in those intense bursts of creative energy that need no further development and leave no foundation to build on beside the memory of their ingenuity. In contrast to those artistic formulae (such as narrative) that can be used over and over again and be adapted to many types of content and effects, the idea that forms the message of conceptual art exhausts its expressive potential after a single use. While it would be easy to reproduce the interface of *The Jew's Daughter, Chemical Landscape,* or *Translation,* the operation would be pointless, even if the text were given different content, because the idea has already been implemented.[8] That the idea cannot be reused also means that once the audience gets the idea, there is little need to explore the text any further. As Umberto Eco writes about Marc Saporta's *Composition No. 1,* a dysfunctional narrative written on a deck of cards that can be freely shuffled, "A brief look at the book was enough to tell me what its mechanism was, and what vision of life (and obviously, what vision of literature) it proposed, after which I did not feel the slightest desire to read even one of its loose pages, despite its promise to yield a different story every time it was shuffled. To me, the book had exhausted all its possible readings in the very enunciation of its constructive idea" (*Open Work,* 170–71). In contrast to texts that create worlds, conceptual texts are not supposed to be read in their entirety but rather to be sampled in quick grabs, like the output of a Web cam. Since the entire text is the product of one generative idea, this idea is fully represented in each grab.

SEVEN

The Many Forms of Interactivity

We tend to think of interactivity as a phenomenon made possible by computer technology, but it is a dimension of face-to-face interaction that was shut off by manuscript and print writing and introduced into written messages by the electronic medium, together with several other features of oral communication, such as real-time (synchronous) exchange, spontaneity of expression, and volatility of inscription.[1] A *New Yorker* cartoon reminds us of the interactive character of oral storytelling. In it, a man sits at his daughter's bedside with an open book in his hands. "Stop asking so many questions," exclaims the exasperated father, "or it's right back to Books on Tape for you" (June 22–29, 1998, p. 93). From the father's viewpoint, the daughter's many questions are annoying interruptions that lengthen the chore of evening storytelling, but from the daughter's viewpoint, besides retarding the daily banishment to the land of sleep, questions and requests are a way to customize the text to her own desires. Thanks to the inherently interactive nature of oral storytelling, the child is able to elicit explanations, to get her father to develop or retell favorite passages, to encourage him to embellish the descriptions of certain objects, or to force him to change direction if he embarks on a track that she does not approve of. (Children, notoriously, do not like variations on the "real" story.)

The term interactivity is widely used in common parlance and in advertising (I once bought an "interactive cat toy"), but some scholars find it too vague. Janet Murray (*Hamlet,* 128) prefers *agency* to *interactivity,* associating the former with purposeful actions that alter a world meaningfully, in contrast to the latter, which may consist of routine actions such as rolling dice or clicking on links. Espen Aarseth, who claims that "[t]he word interactive . . . connotes various vague ideas of computer screens, user freedom, and personalized media, while denoting nothing" (*Cybertext,* 48), proposes the concept of ergodic to describe texts commonly viewed as interactive. According to Aarseth's definition, ergodic literature is a class of works in which "non-trivial effort is required to allow the reader to traverse the text" (1). In this line of thought, turning the pages of a book is a trivial effort, but clicking on links while surfing the Web or reading hypertext fiction is not because these activities involve choice. But if we understand ergodic design as a built-in mechanism or feed-back loop that enables a textual display to modify itself, so that users will encounter different sequences of signs during different sessions, then the "non-trivial effort" does not necessarily come from the user. For instance, "The Speaking Clock," a digital poem by John Cayley, generates an ever-changing display of words selected from a fixed textual database, using the computer clock, as well as aleatory selection devices, to trigger transformations. Since ergodic texts may be closed systems that operate without human intervention, I believe that ergodism and interactivity are potentially distinct properties, even though they often fall together, and that interactivity as human intervention retains a legitimate place in the toolbox of media studies. There are noninteractive ergodic texts, such as "The Speaking Clock," or the output of an Eolian flute operated by the wind; nonergodic interactive texts, such as conversation (which, as free-flowing exchange, lacks the global design of ergodism); and texts that are both ergodic and interactive, such as hypertext and computer games. Moreover, neither ergodism nor interactivity presupposes a digital platform, as the examples of conversation and of oral storytelling (interactive, not ergodic) and of the Eolian flute (ergodic, not interactive) suggest.

Interactivity appears on two levels: one constituted by the medium, or technological support, the other intrinsic to the work itself. All interactive works necessitate a reasonably interactive medium, but interactive media can transmit noninteractive works. Television, for instance, is a mildly interactive medium, since it enables users to switch channels at will (even more conveniently since the invention of the remote-control device), but this feature only allows a choice among noninteractive programs. Interactivity of the medium is detrimental in this case to the appreciation of the individual texts, since by channel surfing the spectator violates their putative integrity. Similarly, the Internet as a whole is an interactive medium, but many of the documents it makes available are themselves standard linear texts. The inherent interactivity of digital media has however fostered a major phenomenon that Henry Jenkins (*Convergence Culture*) calls participatory culture. By connecting people with similar interests and by providing easy-to-use tools that turn consumers into potential creators of content, digital media have tremendously facilitated or even made altogether possible behaviors such as creating fan fiction that expands, modifies, or satirizes the world of popular narratives, writing and publishing book reviews on platforms such as Amazon, exchanging opinions on blogs, commenting on news on public fora, e-mailing the characters or actors of online soap operas, and collaborating with other fans to solve the riddles of Alternate Reality games. Most of these activities fall outside the scope of this book, however, because they develop around noninteractive works, such as popular novels, films, or TV series.

Types of Interactivity

In contrast to the behaviors that make up participatory culture, what I regard in this chapter as interactivity is a mechanism inherent to the design of texts, especially of narrative texts. In my book *Avatars of Story* (107–22; see also Aarseth, *Cybertext,* 62–67), I propose a taxonomy of interactive devices that relies on two dichotomies: internal versus external to the storyworld, and ontological versus exploratory. Here I summarize the argument.

Internal versus External Interactivity

When the user of an interactive text plays the role of an individuated member of the storyworld, interactivity is internal. The actions of the user correspond to events in the history of the world. On the other hand, when she does not imagine herself as a particular member of the storyworld, or when she controls the storyworld from a godlike perspective, interactivity is external.

Exploratory versus Ontological Interactivity

In exploratory interactivity, the user looks at what exists in the storyworld but has no creative power. Her involvement with the storyworld has no lasting consequences. In the ontological variant, her actions create objects that become part of the storyworld or cause events that bring lasting changes. The storyworld evolves as a result of the interaction. While exploratory interactivity precludes world-changing actions, ontological interactivity is a broader domain that allows purely explorative actions.

These two dichotomies can be cross classified into four types of interactivity.

External-exploratory

This is the situation found in the classical hypertext narratives of the 1990s, such as Michael Joyce's *afternoon,* Stuart Moulthrop's *Victory Garden,* or Shelley Jackson's *Patchwork Girl.* "In the texts of these groups, the user is external to both the time and space of the virtual world. There are no time limits to the user's actions;

these actions do not simulate the behavior of a member of the virtual world; and interactivity is limited to the freedom to choose routes in a textual space that has nothing to do with the physical space of a narrative setting" (Ryan, *Avatars*, 108). It could be argued that by making a choice among a set of links that lead to different plot developments, the user determines the fate of the characters and therefore exercises ontological power over the storyworld (Bode, *Future Narratives*, 52–55). This situation would occur if it were impossible to return to the same decision point and to take the other branch—if, in other words, the path not taken were pruned from the set of possible developments and the hypertextual network were turned into a tree (cf. the Julian Barnes quote in chapter 8). Insofar as readers can loop back and follow other possibilities, however, their involvement with the storyworld is exploratory rather than ontological. They look at the various possible worlds that have been encoded in the network, but their choice does not actualize one of these worlds at the expense of the others; rather, the purpose of interactivity is to probe the field of the possible for its own sake.[2] It is also in the category of the external-exploratory that I place interactive digital works that do not create a storyworld—in other words, that are not narrative—such as remixes, aleatory products of generative algorithms, and concrete poetry.

External-ontological

The prime examples of this type of interactivity are the so-called simulation games, such as The Sims or Civilization, in which the user creates the storyworld from a remote, godlike perspective, typically by selecting elements from a menu. Since the player controls several entities, he does not identify with any of them but rather plays the role of a puppet master who holds the strings of characters, choosing their properties and making decisions for them. The system computes the consequences of these decisions and generates sequences of events that may surprise the player, for the puppet master or god figure is not omnipotent. Just as, in real life, the outcome of our actions is determined by countless forces outside our control, in external-ontological interactivity the evolution of the storyworld is determined by the compatibility of the actions selected by the player with the total state of the world. The same action, as represented by a menu choice, will lead to different consequences depending on the particular circumstances.

Internal-ontological

This form of interactivity is the most common in computer games. It is represented by first-person shooters and by adventure games, but its strongest manifestations are, in the nondigital domain, tabletop role-playing games and, in the digital domain, the online world Second Life. The system (which can be a computer program or a human game master) projects a storyworld, and the user impersonates and sometimes creates an individuated member of this world. She may also be able to build objects that become a permanent part of the storyworld. Whether she interacts with NPCs (nonplaying characters), as in computer games, or with real human beings, as in role-playing games and Second Life, the user "writes" the life (or lives) of her character and, by extension, the history of the entire world, since a world's history is the sum of the personal histories of all of its members. The narrative is created dramatically, by being enacted, rather than diegetically, by representing past events. Because of the phenomenological similarity between living one's life in the real world and living one's avatar life in a virtual world, the narrativity of this mode of involvement has been called into question by both narratologists (Porter Abbott) and game theorists (Aarseth, Juul, Eskelinen). It can be argued that "living one's life" is an unmediated experience of the world, while narrative, as a representation, provides a mediated experience. Yet when players recount their adventures and their exploits in a narratively designed game world, their discourse invariably takes the form of a story. This suggests that their experience of the game world was a narrative one. Similarly, life has a narrative potential that people actualize when they represent their personal experiences from a retrospective point of view.

Internal-exploratory

While logically possible, this combination is uncommon, at least in a pure form. The user experiences the storyworld from the inside, that is, from a perspective that reflects the embodied point of view of one of its members, and she is able to move around the world, play tourist, pick up objects, look at them, even alter her perspective, but she has no power to change anything. An excellent example of this mode is a game called The Manhole (1988) by Robyn and Randy Miller, the authors of the famous Myst series. The Manhole consists of a series of still pictures activated by clicking on invisible hot spots. While the user receives no specific identity in the storyworld, the pictures are seen from the vantage of a first-person point of view (the point of view made popular by those shooting games in which users can only see the parts of their body that one can see in real life), and passing from one picture to the next gives the viewer the impression of looking out from a moving body whose altered perception reflects a change of position. In contrast to standard computer games, The Manhole has no puzzles, and the appeal of the game lies entirely in the exploration of the world's spatial design. The internal-exploratory mode is also manifested in art installations in which digital technology tracks the movements of the interactor's body and modifies the display accordingly.

Interactive Structures

The narrative potential of an interactive text is a function of its underlying structure. Figures 8 through 17 illustrate interactive architectures that support various types of narratives or antinarratives.

The Vector (with Optional Side Branches)

This architecture is the most compatible with narrative, since it preserves the linearity of its temporal sequence and of its causal structure. But since there is no other choice than moving forward, can a vector-based narrative be considered ergodic, and can it support interactivity? An example of a digital narrative based on this structure is Kate Pullinger's charming *Inanimate Alice.* Using text, photos, animation, and sound, this work follows the adventures of a little girl who moves to different places—northern China, Italy, Moscow, central England—every time her father takes a new job. Each episode consists of a series of pictures that can be explored by clicking and moving the cursor. Fully exploiting the multimodal capabilities of the medium, these actions alter the picture, launch sounds, trigger animations, or bring more text to the screen. Sometimes it is necessary to complete a game, such as running a maze, to move to the next unit, but most of the time the story is strictly linear, and the user can progress by simply following arrows. Except for the game parts, which are few and appear seldom, this narrative could easily unfold by itself like a film. What, then, is the purpose of the interactive mechanism? By waiting for the user's response before moving to the next screen, the interface enables her to read the text at her own pace. I also believe that because the activity is effortless and the path forward well marked, the effect of having to click in order to get more of the text or to see new places is to engage the reader more deeply with the storyworld and to strengthen her emotional bond with Alice, who, like the reader, is constantly trying to adapt to new environments. Clicking on links (or nowadays touching a screen) to get more information may have become so familiar that it can no longer be considered a "nontrivial effort," but there is still something magic about summoning new information to the screen as a result of physical action—magic because it takes hidden, mysterious code to mediate between the gesture of the hand and its visible effect.

A variation on the vector structure that makes it more interactive is to include optional side branches that lead to roadside attractions (figure 8). In some versions, the user has to explore each screen thoroughly before being allowed to continue; in others, she can progress without visiting all the attractions. This structure is particularly popular in electronic texts meant for juvenile audiences because of its cognitive simplicity. In *Arthur's Teacher Troubles,* a children's book produced by Brøderbund, the user moves page by page through an illustrated story, but every page offers hidden surprises. Click on the teacher, and she turns into a monster; click on one of the students, and he makes a funny face; click on a mirror on the wall, and it is shattered by a baseball; click on the text under the picture, and it is read aloud. We also find this structure in digitally augmented books, such as Marisha Pessl's *Night Film.* The novel, published as a print book, is a multimodal text that combines a linear written narrative with many illustrations corresponding to documents, such as illustrated magazine articles, newspaper reports, and especially Web pages. Some of the illustrations can be scanned with an electronic device such as a smart phone or tablet computer, revealing online content of a semiotic nature that could not fit in a book, such as an audio recording of a character playing the piano, or providing documents that would take too much space in the primary text, such as a detailed syllabus of a course taught by a character. This additional content expands the storyworld without affecting the logic of the plot, so that readers can take either a short route through the book or a long route that follows the sides branches.

The Complete Graph

When it comes to freedom of choice, the polar opposite of the vector is the complete graph (figure 9). In this architecture, every node is linked to every other node, and the reader has total freedom of navigation. This structure allows a free shuffling of text that makes it practically impossible to guarantee narrative coherence. To produce a collection of freely combinable lexias that would generate for every possible order a well-constructed and different story would be a feat of the same order and of the same interest—mathematical rather than poetic—as composing a large crossword puzzle without black spaces, filling a square with numbers that add to the same sum in all rows, columns, and diagonals, or designing a jigsaw puzzle whose pieces interlock in several visual patterns, each corresponding to the realistic image of a different object. Needless to say, the rare examples of this structure (first implemented in Marc Saporta's *Composition No 1*), fall short of the ideal of magic narrative square.

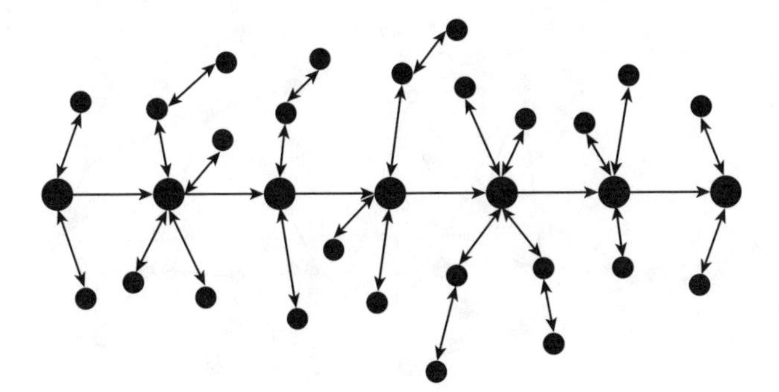

Figure 8. The vector with side branches

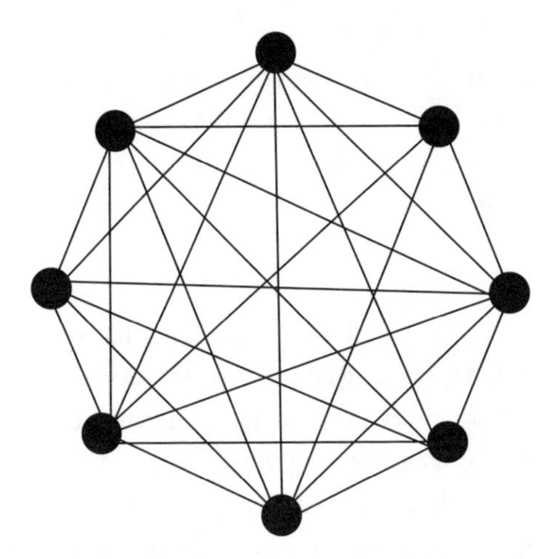

Figure 9. The complete graph

The Network

Figure 10 is the standard structure of literary hypertext. In this architecture, the reader's movements are neither completely free, since they depend on links, nor limited to a single course. The formal characteristic of a network is that it contains circuits and consequently allows nodes to be accessed through different routes. This makes it impossible for the author to control the duration or the course of the user's visit. In such a configuration, narrative continuity can only be guaranteed on the local level (i.e., from one node to the next or within a sequence of nodes with single connections). A reader may, for instance, traverse a node that describes the death of a character and return later to a node in which she is still alive. If the path of navigation is interpreted as chronological succession, the result will be a

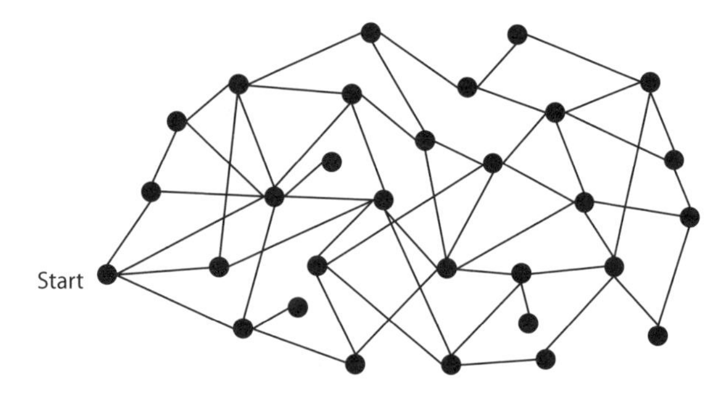

Figure 10. The network: A hypertext-style decision map allowing circuits

sequence that violates the principles of narrative coherence. The network architecture is therefore better suited for a system of analogical connections, for a Dadaist/Surrealist carnivalization of meaning, or for the explorations of multiple alternatives than for the generation of coherent stories, since stories are fundamentally linear sequences of events.[3]

The Tree

The formal characteristic of an arborescent graph (figure 11) is that it allows no circuits. Once a branch has been taken, there is no possible return to the decision point, and there is only one way to reach a given terminal node. By keeping each of its branches strictly isolated from the others, tree-shaped diagrams such as the graph of figure 11 control the reader's itinerary from root node to leaf nodes, and they make it easy to guarantee that choices will always result in a well-formed story. They are, therefore, the structure of predilection of the *Choose Your Own Adventures* children stories. For the sake of graphic simplicity, I have represented the choices as binary, but we can imagine a more interactive tree offering a wide range of possible actions at every decision point. Since the tree grows exponentially, this system quickly runs into a combinatorial explosion. It would, for instance, take sixteen different plots, with thirty-one different fragments, to ensure four decision points. The number of fragments necessary to produce a certain number of plots can be restricted by allowing the merging of paths (dotted lines), but this structure is a directed graph (cf. figure 14) and no longer a tree.[4]

The Database

By database, I mean the structure typical of informational Web sites. Users enter these sites through a home page that offers a menu of different options. As they

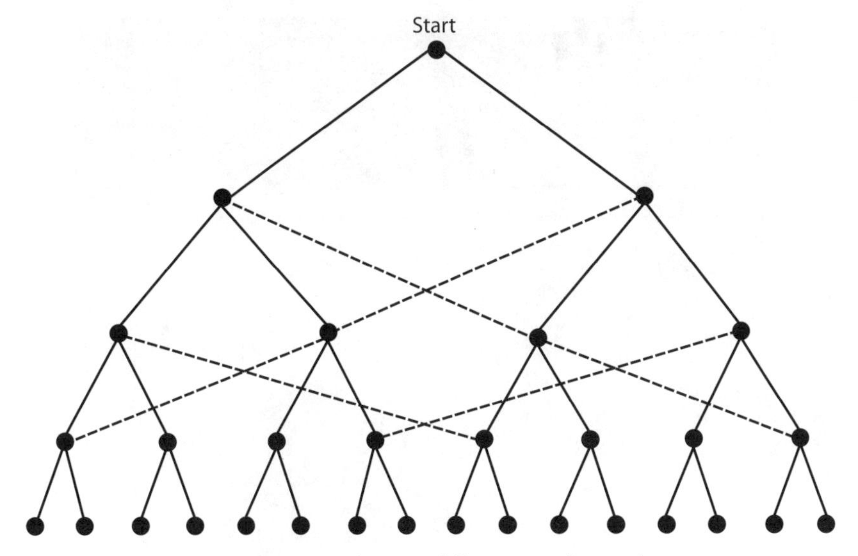

Figure 11. The tree

select one of these options and plunge deeper into the Web site (for it has distinct levels), the original menu remains displayed at the top or on the side of the page, so that you can return in one click to one of the main menu options, rather than having to backtrack. This means that every node of the system is connected to the items of the main menu. This structure, which combines the hierarchical organization of trees with the complex connectivity of networks, makes the exploration of the system much more content-driven than the configuration of figure 10, which usually supports blind progression. A narrative example of this structure is *Marble Springs,* a text of hypertext fiction by Deena Larsen that tells the stories of all the inhabitants of a fictional mining town in Colorado in the nineteenth century (figure 12). Even before there were social media, people were connected into multiple networks: through family relations, through work, through neighbors, through accidental events, through alliances and feuds, through love affairs and enduring grudges. *Marble Springs* uses these networks to provide multiple accesses to characters. The reader can reach them through their family tree, through their grave in the cemetery, through the location of their house or workplace on the map of the town, through an index that lists the various types of stories (Accidents, Adultery, Asylums, Battered Women, etc.), and through the stories that involve several

Figure 12. The database: *Marble Springs*, by Deena Larsen

participants. Within each story the reader can click on all the names of people and places to get background information. (The places function as characters as much as the people do.) But no matter what page the reader is currently visiting, the items of the main menu remain accessible. Two of the main menu tabs, "Humans" and "Maps and Places," lead into the storyworld; the others lead to paratexts that tell about the work or that invite readers to contribute stories. Through "Humans" and "Maps and Places," the reader gets lists of characters and locations that allow random access to the stories relating to these entities. It is this possibility of random access that makes *Marble Springs* a database narrative, though "randomness" does not mean blind progression on the part of the reader but, on the contrary, opens the possibility to deliberately select a given character or place from anywhere in the network. (I say more on database and narrativity in the next chapter.)

The Maze

This structure, shown in figure 13, is characteristic of adventure and shooter games. It differs from the network in that it has one or more exits, which the user tries to find. This scheme has room for many variations. There may be one or more ways

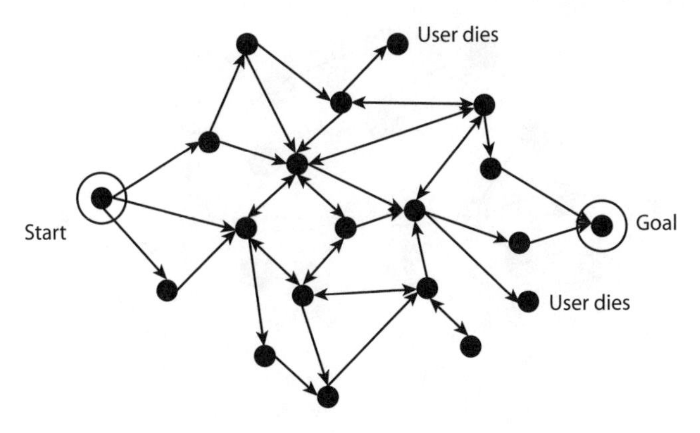

Figure 13. The maze: Structure of an adventure game

to reach the goal; the graph may or may not allow the user to run in circles; terminal nodes may be dead-ends or allow backtracking. Far from guaranteeing satisfaction for every traversal, the system structures the reader-player's adventure in the storyworld in terms of the two experiences that Espen Aarseth calls *aporia* and *epiphany* (cf. Aarseth's essay "Aporia and Epiphany"). While *aporia* occurs when the player takes a dead-end branch on the game-map or fails to overcome an obstacle, *epiphany* is a discovery, such as the solving of an enigma or the elimination of an opponent, that enables him to progress in his quest. Though the structure of the text is too complex for the designer to foresee every possible path, narrative coherence is guaranteed by the fact that all paths are attempts to reach a certain goal. The model contains as many plots as there are complete traversals of the graph. Plots might have a happy ending or tragically end with the waste of one of the user's lives, depending on whether the path ends with epiphany or fatal aporia. But by saving the game before attempting a dangerous action, players can avoid starting all over again.

The Flowchart

The negative experiences of running in circles or hitting a dead-end are eliminated in the architecture of figure 14. This model represents the best way to reconcile a reasonably dramatic narrative with some degree of interactivity. In this type of network, horizontal progression corresponds to chronological sequence, while the branches superposed on the vertical axis represent the choices offered to the user. The system prescribes an itinerary through the storyworld, but the user is granted some freedom in connecting the various stages of his journey. In one variant of the structure (illustrated by the novel *The French Lieutenant's Woman*), the reader

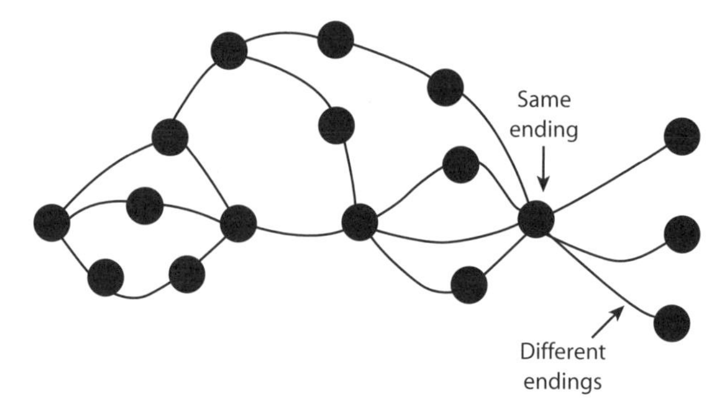

Figure 14. The directed network, or flow chart

receives a choice of endings at the conclusion of a unilinear journey. This architecture prevents the combinatorial explosion of the tree-shaped diagram, but when the choices occur in the middle exclusively, it trivializes the consequences of the user's decisions. As Gareth Rees observes in "Tree Fiction on the World Wide Web," "the merging of narratives keep[s] the story on a single track while offering [the user] an illusion of choice." If the user can get from A to B by different paths and from B move on to C, the choice among the various paths bears no consequence for the final outcome.

One way to restore significance to the user's decisions is to turn the text from a fully context-free to a context-sensitive transition system capable of narrative memory. In such a system, the decisions made by the user in the past affect his choices in the future, and narrative causality extends to nonadjacent episodes. An example of this distant causality is provided by the classical Proppian fairy tale. Imagine that the hero, on his way to rescue the princess from the dragon, is tested by a donor. He may either pass the test and receive a magical aid or fail and be punished with a curse. When he encounters the dragon later in the story, the outcome of the fight is determined by whether or not he carries the donor's gift. Many computer games implement this idea by having players pick up and carry objects that will enable them to solve later problems. This use of memory makes it possible to include nontrivial choices at every stage in the story and to make the end dependent on the middle. The user, however, is not necessarily aware of how his decisions affect narrative development. The game Heavy Rain, for instance, has multiple endings, all determined by the player's actions, but the causality of the relation is not a matter of narrative logic accessible through common sense reasoning (such as the player realizing that he cannot kill the dragon because he missed the chance to acquire proper weapons) but the product of fortuitous coincidence determined by multiple variables: if conditions a, b, and c obtain, the code implements ending 1; if a, b, and d obtain, the code implements 2, and so on.

The Hidden Story

Figure 15 is the structure of those interactive mystery stories and computer games that implement the idea of discovering the prehistory of the game world, as in the popular game Myst. This model consists of two narrative levels: at the bottom, the fixed, unilinear, temporally directed story of the events to be reconstituted; on top, the atemporal network of choices that determines the reader-detective's investigation of the case. Between the two are dotted lines that link episodes of discovery in the top story to the discovered facts of the bottom story. In this case, as in the maze configuration of figure 10, a narrative is written by the player's actions and movements in the attempt to reconstitute the underlying story.

The Braided Plot, or, the House of Many Windows

Classical narrative consists, at least in part, of a sequence of physical events objectively experienced by a group of characters, but every character in the cast lives these events from a different perspective and has a different story to tell. Figure 16 shows how interactive mechanisms can be used to switch "windows" on a multistranded but determinate narrative. On the diagram, the horizontal axis stands for time and the vertical axis for space; simultaneous events are vertically aligned, and events that take place in the same location occupy the same horizontal coordinate. Each circle represents a physical event, and the lines that connect them stand for the destinies of the participants. By selecting a horizontal line rather than another, the reader enters the private worlds of a specific character and experiences the story from a particular point of view. This architecture, like many others, allows a number of variations. Some systems may restrict switching between plot lines to certain points or, on the contrary, make switches possible at every moment. Some may let the reader backtrack and relive the same events from a different point of view, while others may impose a forward movement, so that the perspectives not taken will be irrevocably lost. An example of unconstrained movement among the threads of different characters is M. D. Coverley's hypertext *Califia.* The text chronicles a treasure hunt in California narrated by a character named Augusta. She is accompanied in her quest by two other characters, Kaye, the mystic, who gathers star charts, Indian lore, and various spiritual charts, and Calvin, the archivist and media specialist, who maintains an archive of documents about California

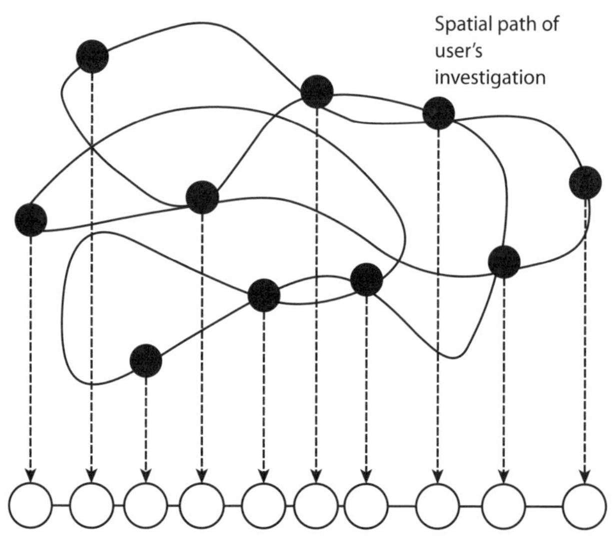

Figure 15. The hidden story

and family history, including diaries, letters, photos, and maps. The hypertextual network is connected in such a way that it is always possible (or possible with minimal action) to pass from the Augusta to the Kaye or to the Calvin threads, and vice versa. A (nonelectronic) example of lateral movement that leads to a loss of information concerning other branches would be a play whose action unfolds simultaneously in several rooms of the same house. The spectator would be able to move from room to room, but she could not view what happens at the same time in other rooms, and since time flows irrevocably forward, she could not backtrack to see what she has missed. It would therefore take the viewing of many different performances to catch the whole story.[5]

Action-Space

In the architecture of figure 14, the system designs the general outline of the plot, and the user selects the details of its realization. Figure 17 represents the inverse solution. Here, interactivity takes place on the macro-level and narrative plotting on the micro-level. The space of the diagram represents the geography of the virtual world, and the nodes and links correspond to the prominent sites and access ways in this geography. The user is free to take any road, but when she reaches a site, the system takes control of her fate and sends her into a self-contained adventure

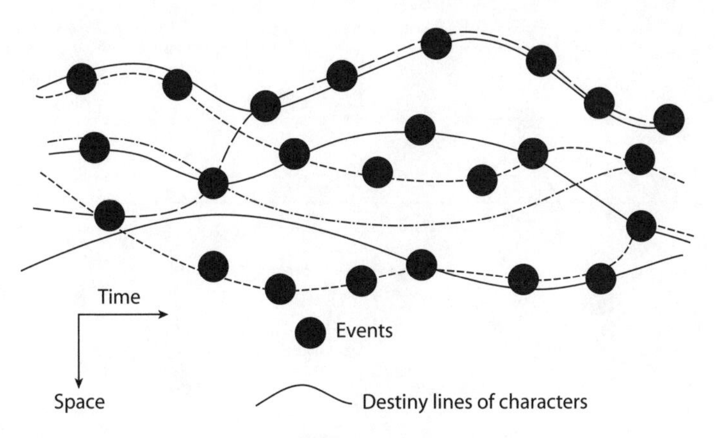

Figure 16. The braided plot: The house of many windows

(represented by the pathways looping back toward the nodes). This model abandons the idea of an overarching dramatic narrative in favor of an epic structure of semi-autonomous episodes.

This model is illustrated by the structure of theme parks. Visitors wander in a geography made of distinct subworlds, each of which offers a different, carefully scripted adventure. In one site, you take a boat through the pirates' lair; in another, you ascend the Matterhorn in a train; in a third, you venture on a rocket into intergalactic space. Choice is limited to deciding where to go and whether or not to take the ride; once the ride has been boarded, the system takes total control of the visitor's fate and (hopefully!) provides a thrilling experience with proper dramatic contour. We can find another example of this architecture, which Henry Jenkins ("Game Design") calls spatial storytelling, in MMORPGs (massively multiplayer online role-playing games) such as World of Warcraft. In the design of online worlds, there is no overarching story but countless little stories that relate to the various regions of the world's geography. For instance, if you wander near a certain village, you will meet a NPC who will gossip about the people of the village. Further down the road, another character will tell you about a serious problem that plagues the area and will give you a quest to perform.[6] Quests consist typically of a backstory told by the NPC that explains the need for action and of a sequence of predefined steps that, when performed by the player, will complete the story.

Levels of Interactivity

Another way to categorize forms of interactivity is to base the taxonomy on the level of the text affected by the user's interventions. Narrative texts, whatever their

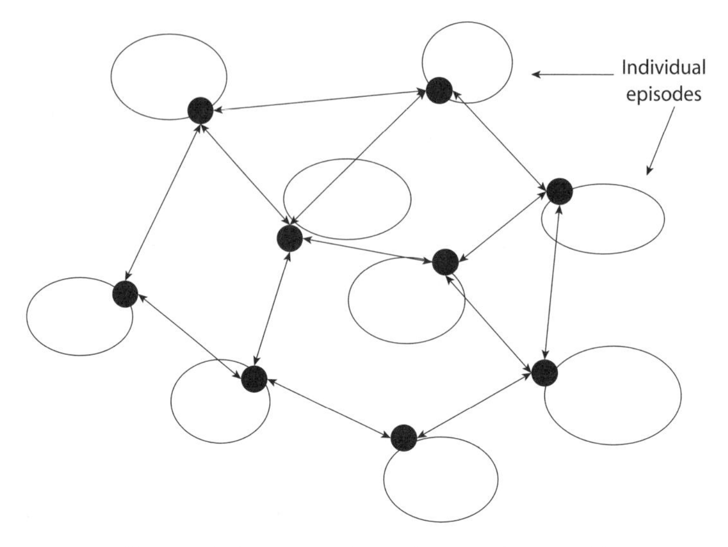

Figure 17. Action space, epic wandering, and storyworld

material embodiment, are layered like an onion, with the outer layers made of discourse and the inner layers made of story. In this section, I propose to peel this onion, discussing forms of interactivity that relate to its various levels, as well as the works that exemplify these forms.

Level 1. Peripheral Interactivity

Here the story (or more generally, the content) is framed by an interactive interface, but the user's activity affects neither the story itself nor the order of the discourse that presents it. Its role is to make visible the signifiers, so as to render the text legible.

My first example is the poem "Cruising," by Ingrid Ankerson and Megan Sapnar (figure 18). The "story" of the poem is a memory of growing up in a small Wisconsin town. It is not a full-fledged narrative because it describes a somewhat repetitive action, and it does not reach any kind of narrative closure. But the reader can relate emotionally to the characters, and the setting sparks the imagination of anybody familiar with American small towns. The user's action consists of controlling the display. By moving the cursor, the reader can make the text and its graphic background grow or shrink, move left or move right, and move at different speeds. The goal is to get a combination of size, speed, and direction that allows the text to be deciphered; for most of the time, the letters are too small, and they move too

Figure 18. Cruising, by Ingrid Ankerson and Megan Sapnar

fast for the eye to make out the words. The user's control of the speed and direction simulates the driving of a car, and, indeed, driving a car is what the text is all about. This creates a nice unity not only of form and content but also of interface. But no matter how fast or slow the car moves, no matter whether it goes forward or backward, it is the same text that rolls before the reader's eyes. Sometimes we can read it, sometimes we cannot, but we cannot stop it, we cannot skip any of its parts, and we cannot change its internal order.

The viewing of the text is turned into a more challenging operation in my next example of interactive interface, *Marginal,* by Stuart Moulthrop (figure 19). The text consist of a collage of story fragments that seem to be cut out from newspaper columns—an obvious allusion to the cutout technique of William Burroughs. But the text is hidden by an opaque cover. A hole in this cover travels on the screen, revealing parts of the stories, but without allowing the user to read them because it moves too fast and too randomly. The interactivity of the text depends on a second hole whose movement can be fully controlled. By walking this hole slowly over the screen, the user is able to choose which part of the hidden text to expose, and by moving it from left to right over a fragment, she is able to read it. When the controllable hole passes over a certain hot spot, the page is replaced by another, but since the user does not know where the hot spot is located, she has only indirect control over this event. Sometimes the text is instantly replaced before the user can read anything, sometimes it stubbornly refuses to give way to the next page. The

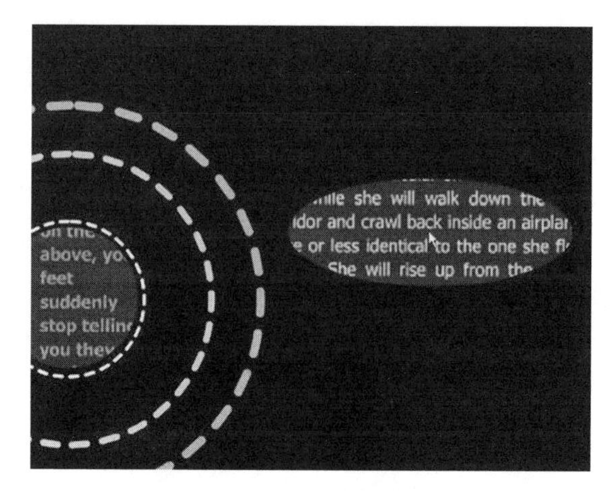

Figure 19. Marginal, by Stuart Moulthrop

effect is like reading a book with a magnifying glass, except that in this case the user is not free to turn the pages. In contrast to "Cruising," however, there isn't any obvious relation between the user's activity and the particular themes of the text (mostly humorous and sexual), and I can't think of any purpose for the interface other than to bring attention to the reading process itself by de-automatizing the scanning of the text by the eye.

Level 2. Interactivity Affecting Narrative Discourse and the Presentation of the Story

On this level, as on the previous one, the materials that constitute the story are fully predetermined, but here the order in which the story is told is highly variable, while in level 1 interactivity affects only the physical readability of the text.

The best examples of discourse-level interactivity are the hypertexts produced in the early 1990s with the program Storyspace, for instance, *afternoon* by Michael Joyce or *Victory Garden* by Stuart Moulthrop. As we have seen above, classic hypertexts are based on a network structure characterized by the existence of loops that offer several different ways to get to the same node. These loops make it possible to circle forever in the network. This explains why the image of the labyrinth and the notion of "book without end" (Yellowlees Douglas) play such an important role in hypertext theory. To reinforce the reader's experience of being lost in a labyrinth, classical hypertext favors opaque links, which lead to random selection and blind navigation. In *afternoon,* links are not visible unless one holds down a certain key. In other texts, they are signaled by underlined words, but in contrast to the links of a database or of a strictly practical Web site, the words themselves have no evident informational value, and the relation between the words that anchor the links and the text that comes to the screen is treated as a puzzle to be solved by the reader. An important dimension of reading hypertext is indeed to ask: Why is this segment linked to this other one?

Hypertext aesthetics favors the serendipitous emergence of meaning over a goal-oriented, deliberate retrieval of information. But is it possible to respect narrative logic under these conditions? Early theorists presented hypertext as a storytelling machine that generates a different narrative with every run of the program. As Michael Joyce puts it: "Every reading . . . becomes a new text. . . . Hypertext narratives become virtual storytellers" (193). But an infinite number of different paths through a network does not necessarily mean an infinite number of different stories, since, according to narrative theory, the same story can be told in many different ways. If every path corresponded to a new story, hypertext fiction would implement the type of interactivity that affects the inner layers—an interactivity that creates stories on the fly, rather than disclosing a preexisting story. (See level 4 below.) For this to happen, the order in which the reader encounters the lexias would have to correspond rigidly to the chronological order of the events narrated in the lexias; for if the lexia could be mentally rearranged by the reader, different paths through the network could be read as the same story. But the loops of the network structure of classical hypertext prevent the interpretation of the sequence of lexias as a faithful image of chronological order.

To see what is wrong with associating the sequence of lexias seen by the reader with the sequence of events in the storyworld, consider the lexia in Joyce's *afternoon* where the narrator witnesses an accident and fears the victims were his ex-wife and son:

> Die?
>
> I felt certain it was them. I recognized her car from that distance, no more than a hundred yards off along the road to the left where she would turn if she were taking him to the Country Day school.
>
> Two men stood near the rear of the grey Buick and a woman in a white dress sprawled on the wide lawn before them, two other men crouching near her. Another, smaller body beyond. (1987)

Now imagine that after reading a certain number of other lexias you return to this scene a second and perhaps a third time. There are at last four different ways to interpret this recurrence.

1. The narrator has seen two, three, or four similar accidents, depending on how many time the reader returns to the lexia. We are in a world where events strangely repeat themselves, as in the movie *Groundhog Day.*
2. The narrator travels back in time and sees the same accident over and over again.
3. The narrator is obsessed with the experience of the accident, and the return to the same lexia stands for the replaying of the scene in the narrator's mind. The text represents the stream of consciousness of the narrator.
4. Return to the scene of the accident is nothing more than a return to the same chunk of text, and it has no significance within the storyworld.

The first two interpretations affect the level of story, but they involve a fantastic or science-fictional element that is totally absent from *afternoon,* and this makes them rather silly. The last two interpretations, which I find much more acceptable, involve the level of discourse: the inner discourse of the narrator in 3 and, more abstractly, the discourse of the text in 4. In both interpretations there is only one accident, and interactivity provides many glimpses of the same scene rather than creating different sequences of physical events within the storyworld. But interpretation 3 comes closer to affecting the level of story because it naturalizes the text as the mental activity of a character who exists within the storyworld. In different runs of the text, the narrator's mind will consequently follow different paths and visit different memories. In interpretation 4, by contrast, the textual mechanisms are no longer interpreted mimetically. Interactivity becomes a game of putting a coherent story back together out of fragments that come to the reader in a variable order, like the pieces of a jigsaw puzzle. (This is discussed more in chapter 8.)

Level 3. Interactivity Creating Variations in a Partly Pre-Defined Story

While on the preceding two levels of interactivity the user's participation was external and exploratory, on this level it is internal and either ontological or exploratory. The user plays the role of a member of the storyworld, and the system grants him some freedom of action, but the purpose of the user's agency is to progress along a fixed storyline, and the system remains in firm control of the narrative trajectory. This type of interactivity is typical of many genres of computer games, such as adventure games, shooters, and mystery-solving games.

The game determines the theme of the mission and the sequence of the tasks, but the player's performance creates variations on this fixed frame. Gordon Calleja has proposed the term *alterbiography* to designate "the ongoing narrative generated during interaction with a game environment" (*In-Game*, 124). Though individual alterbiographies are conditioned by a common narrative design, they differ from each other in how the avatar solves problems, in the success or failure of his quest, and, in the case of failure, in the time and manner of his death. The different ways to implement the same global narrative design can be represented as shown in figure 20. But a global design can also present several branches and many different endings. In this case its underlying structure will look like the flowchart of figure 14. To respect narrative logic, this type of structure should only allow strands to merge when it no longer matters which route the avatar has taken. Merging points are fresh starts that generally correspond to the various levels of the game. But the flowchart structure can present partial merging points, corresponding to the many different ways to achieve a subgoal.

The formula of level 3 has been very successful in computer games, but the reason for this success may lie in the fact that game players are not very discriminating when it comes to narrative. Many hard-core players do not play for the story but for the adrenaline rush of competition and for the thrill of beating the game. As long as they get stunning graphics and their dose of fast action, they are satisfied with the same old storyline clothed in different themes and visual motifs. It will take lots of imagination on the part of game designers to make games worth playing for the sake of the story.

Level 4. Interactivity Leading to Real-Time Story Generation

On level 4, stories are not predetermined but generated on the fly out of data that comes in part from the system and in part from the user. Every run of the program should result in a different story, and the program should therefore be replayable.[7]

To this day, we do not really have a story-generating system sufficiently sophisticated to produce a wide variety of interesting stories out of data internal to the system. Integrating the user's input into the generating process only raises the difficulty to a higher power. The major obstacle to the development of truly interactive, generated-on-the-fly narrative is not technological but logical and artistic. How can the freedom of the user be reconciled with the need to produce a well-formed, aesthetically satisfying story? VR researchers Ruth Aylett and Sandy Louchard refer to this problem as the "interactive paradox": "On one hand the author seeks

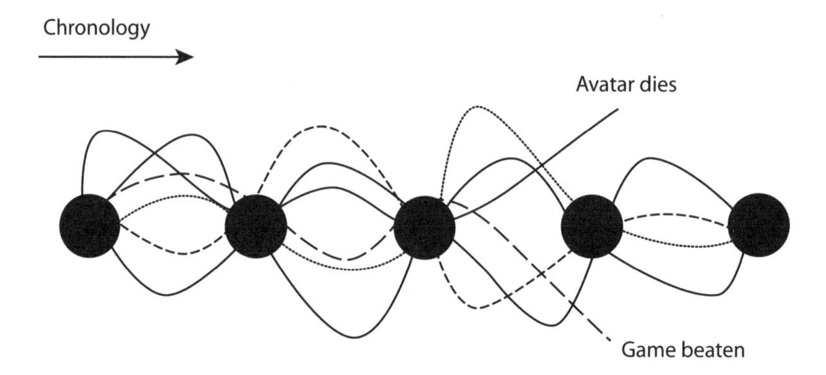

Figure 20. Variations on the same narrative sequence

control over the direction of a narrative in order to give it a satisfactory structure. On the other hand a participating user demands the autonomy to act and react without explicit authorial constraint" ("The Emergent Narrative," 25).

Another way to formulate the paradox is in terms of the discrepancy between the goals of authors and the goals of people engaged in living their own lives. This discrepancy is captured by the formula "Life is lived looking forwards, but it is told looking backwards," attributed to Søren Kierkegaard. When we live our life we ask: What action can I take to solve my problems and reach a more satisfactory state of affairs in the future? But when we tell a story, we start from a situation that we find interesting, and we ask: What course of events led to this situation? The visitor to an interactive narrative system plays the role of a character in a virtual world and adopts the forward-looking perspective of life. When we are faced with a problem in real life, we want to resolve it as quickly and as efficiently as possible. But the author who creates a story is more interested in actions that produce opportunities for interesting plot developments than in efficient problem solving. A particularly telling example of the conflict between character goals and authorial goals is the story of *Little Red Riding Hood.* When the hungry wolf meets the little girl in the forest, why doesn't he eat her on the spot, rather than waiting until she reaches the house of the grandmother? He is taking the risk that Little Red Riding Hood will never find the grandmother's house or that another wolf will eat her in the meantime. But from the perspective of the storyteller, the wolf's plan is infinitely superior to the practical solution because it allows the highly dramatic episode of the wolf tricking the heroine by taking the grandmother's place in bed and the climactic event of their confrontation. In an interactive version of *Little Red Riding*

Hood, should a user playing the wolf take a practical attitude, or should she make a choice that results in higher tellability?

Yet despite the difficulties of combining top-down design with user freedom, many designers believe that without the former, interactive narrative systems would put an excessive burden on the user. The systems must be *authored,* and users should respond to affordances built into the virtual world and programmed into the system, rather than being entirely responsible for constructing the story. Not many of us prefer writing plays and novels to watching and reading them; by the same reasoning, most users of interactive narrative systems prefer being invited into a story than having to create it from ground zero. Nicolas Szilas advocates, for instance, a module he calls the "Virtual Narrator," whose function is to guarantee "storiness" by selecting rules and events on the basis of their effect on the user rather than (exclusively?) on the basis of the behavior of characters ("A New Approach," 3). Aylett and Louchard believe that narrativity in VR systems should "emerge directly from the interactions between the protagonists" rather than from a scripted plot, but they still recommend a "drama manager" function, inspired by the "game master" of tabletop role-playing games, who monitors the story though indirect communication with the players, such as putting nonplaying characters in the players' way to influence their decisions ("The Emergent Narrative").

Level 4 interactivity presupposes an ontological involvement, but this involvement can be either internal or external. The internal stance is found in the interactive drama Façade by Michael Mateas and Andrew Stern (figure 21). In Façade, the user plays the role of a guest who is invited for dinner by a couple of successful young urban professionals, Grace and Trip. During the conversation between the user and his hosts, deep cracks appear in the supposedly perfect marriage of Grace and Trip, and they say many hurtful things to each other. At the end of the evening, the visitor is expulsed from the apartment after one of several possible developments. In most of the endings, the marriage is irremediably broken, but in at least one the player manages to reconcile the warring spouses. This may sound like a pre-scripted narrative similar to those of level 3, especially since all the runs maintain a dramatic development of exposition, conflict, and resolution, but the emergence resides in the details of the dialogue being generated. While different runs of the system may lead to the same ending, the conversation will always be different. (See more discussion of Façade in chapter 10.)

Emergent architecture depends on external participation in the popular computer game The Sims. This kind of system generates stories by creating characters, endowing them with different personalities and possibilities of behaviors, and let-

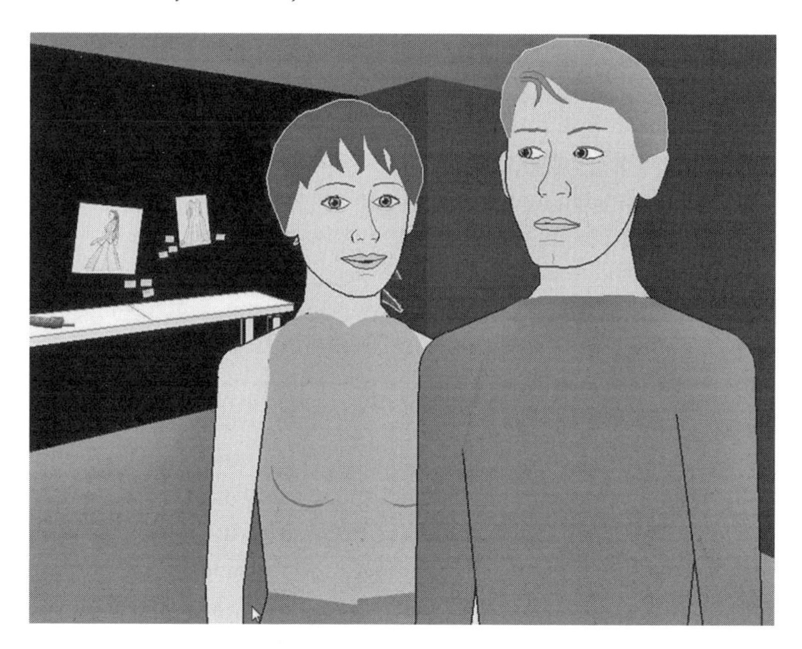

Figure 21. Façade, by Michael Mateas and Andrew Stern

ting the user activate their built-in behaviors. When the user selects an action, the system computes its consequences and updates the current state of the storyworld, opening up a new set of possible actions. Imagine that the user is currently controlling a character called Nina and that Nina's sister, Dina, is present on the scene. By moving the cursor over Dina, the player activates a menu of the various actions that Nina can perform with Dina as a patient: actions such as kiss, hug, dance with, play with, talk to, appreciate, or irritate. All these actions have a certain emotional impact on the patient and affect the relation between the two characters. This relation determines how Dina will respond in the future to Nina's requests. Now imagine that instead of just two characters the world includes many and that the characters can also interact with a variety of objects that affect their mood. The choices offered to the player become too numerous and the interactions of the various objects too complex for the designer to anticipate all the possible developments. The sheer number of possibilities makes this kind of system impossible to diagram; this is why I did not illustrate it in the previous section.

Level 5. Meta-Interactivity

On this level, the interactor is not using the system but, rather, modifying it for other users. Meta-interactivity can take such forms as designing a new level for a computer game, creating new costumes for an avatar, building new objects, associating existent objects with new behaviors, and generally expanding the possibilities of action offered by the storyworld. To constitute a genuinely "meta" interactivity, these actions must be performed by writing code and patching up the source, rather than by using tools internal to the game, such as the house-building module of The Sims or the construction kit of Second Life. It is on this level that the idea of the user as coauthor becomes more than a hyperbolic cliché, but the two roles do not merge, since users cannot simultaneously immerse themselves in a storyworld and write the code that brings this world to life.

The inner layers of the onion are much harder to conquer than the outer layers, but we should not confuse problem-solving difficulty with aesthetic value. There is a tendency in digital culture to evaluate a work as a feat of programming virtuosity. I call this the anti-WYSIWYG aesthetics because you have to imagine the code that lies behind the screen to appreciate the text.[8] By these standards, a work of level 4 is automatically superior to a work of level 1, regardless of its narrative quality, because it requires much more elaborate and original coding. If we applied the same aesthetics to print literature, a palindrome story or a novel written without the letter "e" (such as Georges Perec's *La disparition*) would automatically represent a greater artwork than a novel like Marcel Proust's *A La recherche du temps perdu,* which was written without stringent formal constraints (though certainly not without form). Another aesthetic criterion popular in digital culture that favors the works of the inner layer is the idea of emergence and self-renewability. Façade can be replayed half a dozen times with different results, while the texts of level 3 will rarely be replayed once the game has been beaten. And while it is possible to fiddle for a long time with a hypertext of level 2 like *afternoon,* the works of level 1 quickly yield all of their substance. But for the reader who truly cares for the story, an interactive work that produces a relatively fixed plot but gives intense pleasure during its unique run is not inherently inferior to a system that creates a wide variety of mediocre stories. I am not saying that diversity of output does not contribute positively to aesthetic value but, rather, that a work can compensate for a lack of replayability with other qualities. There are consequently good and bad solutions, success and failure, entertainment and boredom on all the layers of the interactive onion.

EIGHT

Hypertext: The Functions and Effects of Exploratory, External Interactivity

Quain was in the habit of arguing that readers were an almost extinct species. "Every European," he reasoned, "is a writer, potentially or in fact." He also affirmed that of the various pleasures offered by literature, the greatest is invention. Since not everyone is capable of this pleasure, many must content themselves with shams. For these "imperfect writers," whose name is legion, Quain wrote the eight stories in Statements. Each of them prefigures or promises a good plot, deliberately frustrated by the author. One of them—not the best—insinuates two arguments. The reader, led astray by vanity, thinks he has invented them.

—*Jorge Luis Borges*

With the increasing extension of the press, which kept placing new political, religious, scientific, professional, and local organs before the readers, an increasing number of readers became writers. . . . Thus, the distinction between author and public is about to lose its basic character. . . . At any moment the reader is ready to turn into a writer. —*Walter Benjamin*

In the mid-1990s, as digital technology made its way into everyday life, a new, digital-born literary genre appeared on the computer screen. This genre, hypertext fiction,[1] was heralded as the future of narrative, and the novelist Robert Coover, one of the main advocates of electronic writing, saw it as the most significant manifestation of computer-mediated literary art: "And I continue to feel that, for all the wondrous and provocative invasions of text by sound and image, all the intimate layering of them and irresistible fusions, still, the most radical and distinctive literary contribution of the computer has been the multilinear hypertextual webwork of text spaces, or, as one might say, the intimate layering and fusion of imagined spatiality and temporality" ("Literary Hypertext," 19).

Twenty years later, hypertext has firmly established itself as the organizing principle of the Internet, but as a literary form it is no longer at the forefront of experimental writing, and far from achieving popular dominance as a storytelling practice, it now appears "so nineties." In this chapter, I look at the narrative potential of hypertext and at the enthusiasm it initially generated from a perspective that combines observations made in the late 1990s with the hindsight afforded by the twenty years that have elapsed since the heyday of the genre.

The idea of hypertext is so widely known among the readers of this book that its presentation can be limited to a short refresher. In a hypertextual system, text is broken into fragments ("lexias," for George Landow; "textrons," for Espen Aarseth) and stored in a network whose nodes are connected by electronic links. By clicking on a link (usually a highlighted phrase), the reader causes the system to display the contents of a specific node. A fragment typically contains a number of different links, offering the reader a choice of directions to follow. By letting readers determine their own paths of navigation through the database, hypertext promotes what is customarily regarded as a nonlinear mode of reading.

The Myth of Hypertext

Early theorists of hypertext, such as George Landow, Jay David Bolter, Michael Joyce, and Stuart Moulthrop, promoted the new genre in the academic community—an audience generally hostile to technology but also generally open to postmodern theory—as the fulfillment of the ideas of the most influential French theorists of the day. Among these were Barthes, Derrida, Foucault, Kristeva, Deleuze and Guattari, and their adopted "ancestor" Bakhtin. To cite a few particularly telling examples of this rhetoric, Bolter calls hypertext a "vindication of postmodern theory," as if postmodern ideas were the sort of propositions that can be proved true or false ("Literature in the Electronic Writing Space," 24), Richard Lanham ("The Electronic Word," chapter 4) speaks of an "extraordinary convergence" of postmodern thought and electronic textuality,[2] and Ilana Snyder argues that hypertext teaches "deconstructive skills" that readers, supposedly, do not acquire from standard texts (*Hypertext,* 119).[3]

The list of the features of hypertext that support the postmodernist approach is an impressive one. It is headed by Roland Barthes and Julia Kristeva's notion of *intertextuality,* the practice of integrating a variety of foreign discourses within a text through such mechanisms as quotation, commentary, parody, allusion, imitation, ironic transformation, rewrites, and decontextualizing/recontextualizing operations. Whether intertextuality is regarded as a specific aesthetic program or as the basic condition of literary signification, it is hard to deny that electronic linking is an ideal device for the implementation of intertextual relations. Any two texts can be connected, and by clicking on a link, the reader is instantly transported into an intertext. By facilitating the creation of polyvocal structures that integrate different perspectives without forcing the reader to choose between them, hypertext is uniquely suited to express the aesthetic and political ideals of an intellectual community that has elevated the preservation of diversity as one of its fundamental values. The device also favors a typically postmodern approach to writing closely related to what has been described by Levi-Strauss as *bricolage* (*tinkering*, in Sherry Turkle's translation). In this mode of composition, as Turkle describes it (*Life on the Screen*, 50–73), the writer does not adopt a top-down method, starting with a given idea and breaking it down into constituents, but proceeds bottom-up by fitting together reasonably autonomous fragments, the verbal equivalent of *objets trouvés*, into an artifact whose shape and meaning(s) emerge through the linking process. The result is a patchwork, a collage of disparate elements, what Deleuze and Guattari might have called a *machinic assemblage*. As Silvio Gaggi has shown, this broken-up structure, as well as the dynamic reconfiguration of the text with every new reading, proposes a metaphor for the postmodern conception of the subject as a site of multiple, conflicting, and unstable identities.

While hypertext can bring together the heterogeneous, it can also break apart elements traditionally thought to belong together. The dismantling effect of hypertext is one more way to pursue the typically postmodern challenge of the epistemologically suspect coherence, rationality, and closure of narrative structures, one more way to deny the reader the satisfaction of a totalizing interpretation. Hypertext thus becomes the metaphor for a Lyotardian "postmodern condition" in which grand narratives have been replaced by "little stories" or perhaps by no stories at all—just by a discourse reveling in the Derridean performance of an endless deferral of signification. Through its growth in all directions, hypertext implements one of the favorite notions of postmodernism, the conceptual structure that Deleuze and Guattari call a *rhizome*. In a rhizomatic organization, in opposition to the hierarchical tree structures of rhetorical argumentation, the imagination is not constrained by the need to prove a point or to progress toward a goal, and the writer never needs to sacrifice those bursts of inspiration that cannot be integrated into a linear argument.

Building interactivity into the object of a theoretical mystique, the founding fathers of hypertext theory promoted the new genre as an instrument of liberation from some of the most notorious bêtes noires of postmodern thought: linear logic, logocentrism, arborescent hierarchical structures, and repressive forms of power. George Landow writes for instance that hypertext embodies the ideal of a non-hierarchical, decentered, fundamentally democratic political system that promotes "a dialogic mode of collective endeavor" (*Hypertext,* 283): "As long as any reader has the power to enter the system and leave his or her mark, neither the tyranny of the center nor that of the majority can impose itself" (281). Over forty years ago, Roland Barthes identified the figure of the author as one of these oppressive forms of authority from which readers must be liberated: "We know to give writing its future, it is necessary to overthrow the myth [of the author]: the birth of the reader must be at the cost of the death of the Author" ("The Death of the Author," 78). The purpose of new forms of writing (such as what Barthes called "the scriptible") is "to make the reader no longer a consumer but a producer of text" (*S/Z,* 4). For the critics mentioned above, interactivity is just what the structuralist doctor (would have) ordered: "There is no longer one author but two, as reader joins author in the making of the text," writes Bolter ("Literature in the Electronic Writing Space," 37). For Michael Joyce, hypertexts are "read when they are written and written as they are read" (*Of Two Minds,* 192). Or, to quote Landow again, "Electronic linking reconfigures our experience of both author and authorial property, and this reconception of these ideas promises to affect our conceptions of both the authors (and authority) of texts we study and *of ourselves as authors*" (*Hypertext 2.0,* 25; my italics). In *Grammatron,* a hypertextual novel-*cum*-theory that challenges traditional generic distinctions, Mark Amerika takes the cult of interactivity to new extremes by hailing what he calls "hypertextual consciousness" as the advent of a new stage, perhaps the final one, in the political, spiritual, and artistic growth of mankind:

> The teleportation of Hypertextual Consciousness (HTC) through the smooth space of discourse networks creates an environment where conceptions of authorship, self, originality, narrative and commentary take on different meanings. One can now picture a cyborg-narrator creating a discourse network that serves as a distribution point for various lines of flight to pass through and manipulate data linked together by the collective-self. Directing a site (giving birth to a node), will be one way to reconfigure our notion of authorship but in reconfiguring this notion aren't we in effect radically-altering (killing) the author-

> as-self and opening up a more fluid vista of potential-becomings? (Fragment "Teleport")

How seriously can we take the notion of the reader turned author? Anybody who has suffered the pains of writer's block knows the difference between summoning words from one's mind to express or create ideas and clicking on ready-made signs to bring more ready-made signs to the screen. It is not through hypertext fiction that readers literally become authors but, more prosaically, through the forums, blogs, tweets, and self-presentations of social media, as well as through the fan fiction, machinima, and remixes that people create around or out of their favorite works. This activity may be what Amerika has in mind with HTC, though the networking, multimedia encoding, copying, combining, and sharing capabilities of digital media have more to do with it than the mechanisms of hypertext.

Hypertext and Narrativity

To assess the possibility of telling stories through hypertext, let's start with a text often quoted by early theorists as a model of hypertext, Barthes's description of what he calls the "plural text."

> Let us first posit the image of a triumphant plural, unimpoverished by any constraint of representation (of imitation). In this ideal text, the networks are many and interact, without any one of them being able to surpass the rest; this text is a galaxy of signifiers, not a structure of signifieds; it has no beginning; it is reversible; we gain access to it by several entrances, none of which can be authoritatively declared to be the main one; the codes it mobilizes extend *as far as the eye can reach,* they are indeterminable (meaning here is never subject to a principle of determination, unless by throwing dice); the systems of meaning can take over this absolutely plural text, but their number is never closed, based as it is on the infinity of language. (*S/Z,* 5–6)

To this image of a total or "triumphant pluralism," Barthes opposes a structure that limits the infinity of language and is therefore typical of the "classical" or "readerly" text. This structure is narrativity: "All of which comes down to saying that for the plural text, there cannot be a narrative structure, a grammar or a logic; thus, if one or another of these are sometimes permitted to come forward, it is *in proportion* (giving this expression its full quantitative value) as we are dealing with incompletely plural texts, texts whose plural is more or less parsimonious" (6).

Why cannot the plural text have a narrative structure? Following Barthes's image of the network [*réseau*], we can represent this pluralism as a distributed network. In contrast to a line, a network requires two dimensions: it is therefore a spatial object. The fundamental property of a network, compared to a tree diagram, a flowchart or a vector, is that it has no root node and it allows loops. There is consequently no entry point, and there are many different ways to reach a given node. Once they enter the network, explorers must take a decision every time they reach a node that is connected to many other nodes. It is this constant need to make decisions that elevates, in Barthes's view, the reader of plural texts from a passive consumer to an active producer. Since there is a virtually infinite variety of routes through the network, which means an infinite number of potential interpretations, it is impossible to produce a reading that exhausts the meaning of the text.

Narrativity conflicts with Barthes's vision of a triumphal pluralism for at least three reasons. First, Barthes regards unrestricted pluralism as incompatible with "imitation," or the constraints of representation. When readers regard a group of words as the evocation of a particular object, they reduce the polysemy inherent to each word by focusing on the meaning that relates to the kind of object that is being described, at the expense of the other meanings conventionally encoded in that word. Since narrative is a fundamentally mimetic form of discourse, it cannot avoid this limitation of potential meanings. Poetry, being far less mimetic, is much more respectful of polysemy. Barthes was indeed writing in a time when poetry was regarded as the model of how literature should mean, and though he was never particularly interested in the lyric, he was certainly supportive of theories that regarded the practice of literary art as letting language speak for itself by liberating the multiple meanings inherent in each word.

Second, while a network is a spatial structure, narrativity is a fundamentally temporal structure. This is not to say that space does not play a role in narrative. Stories concern a world that functions as a spatial container for existents; thanks to the actions of these existents, the world undergoes changes. Following a story means building a mental simulation of the changes that take place in a world and of the processes that occasion these changes. Bakhtin expressed this inseparability of space and time through the concept of the chronotope. But space and time have fundamentally different properties. Because space has more than one dimension, one can go from one point in space to another through several routes. But time is one-dimensional; it is traditionally represented by images like a river or the flight of an arrow, which suggest a unidirectional movement along a line. If we represent a story through a diagram, there will be only one trajectory that goes from the world of time 1 to the world of time 2; changing the order of the events will result, in the best cases, in a different story or, in the worst case, in no story at all. When a narrative sequence involves causality, it cannot be inverted because an inversion would mean that the effect precedes the cause—an order that most philosophers regard as logically impossible (e.g., Tooley). Moreover, if time could be represented through a spatial network, the loops of the network would mean that it is possible to return to an earlier time—a feature only found in time-travel stories.

The network structure of the plural text leads to the third reason why Barthes's vision of triumphant pluralism cannot take narrative form. Narrative, as Aristotle taught us, has a beginning and end, but networks, because of their loops, allow endless wandering. Barthes is dreaming of a text that renews itself constantly and holds the reader forever fascinated—which means forever captive—in stark contrast to the consumerist page-turners that we read for the plot, driven by the desire to find out how it ends.

Another argument that questions the storytelling ability of hypertext is Lev Manovich's contrast between database and narrative. Manovich defines database as "a structured collection of data . . . organized for fast search and retrieval by a computer." It is therefore "anything but a simple collection of items" (*The Language of New Media,* 218). Databases are typically organized as tables with rows and columns; the rows correspond to individual items, and the columns to their attributes. This organization allows random access. For instance, in a database that lists various properties of the fifty states of the United States, one can, in one probe, access "Wyoming (a row)-capital of (a column)" and retrieve "Cheyenne." According to Manovich, database organization is incompatible with narrativity: "Many new media objects do not tell stories; they do not have a beginning or end; in fact, they do not have any development, thematically, formally, or otherwise that would organize their elements into a sequence. Instead they are collections of individual items, with every item possessing the same significance as any other" (218). He concludes: "database and narrative are natural enemies" (225). While databases "represents the world as a list of items," narrative "creates a cause and effect trajectory of seemingly unordered items (events)."[4]

Can hypertext be regarded as a database? Yes and no. Hypertexts are not stored as tables with labelled rows and columns but as networks of fragments connected to each other by "goto" statements. This structure does not allow random access; one must follow a sequence of links to reach a given lexia. But in a system such as Storyspace, in which most of the hypertexts of the 1990s were written, the interface gives access to an alphabetical list of all the lexias. This list allows the user to bypass the system of links and to reach any lexia by clicking on its name.[5] Another database feature of hypertext resides in the fact that it is meant to be probed, rather than traversed in a prescribed order, from beginning to end. Readers hardly ever exhaust all the links and nodes, and different readers will see different subsets of lexias in a different order. Yet the vast majority of hypertext fictions are not collections of individual items; readers must visit a reasonable subset of lexias and build a global mental representation (such as a fictional world) out of these elements. Hypertext lies therefore halfway between the spatiality of databases, out of which elements are pulled individually, and the temporality of classical narrative, where meaning arises from a sequential organization.

The texts that are now regarded as the classics of the genre all retain some kind of narrative structure. I am thinking here of texts such as Michael Joyce's *afternoon: a story* (1987) and *Twelve Blue* (online, no date), Stuart Moulthrop's *Victory Garden* (1991), and Shelley Jackson's *Patchwork Girl* (1995). These texts tell about characters who are engaged in certain situations, who perform actions, who interact with other characters, who experience changes in their worlds, and who are emotionally affected by these changes. For hypertext to retain some degree of narrativity, there must be limitations on the connectivity of the network. By this I mean that there must be reasonably long stretches of nodes with a linear connection so that they can be interpreted as a chronological and causal sequence. But for the hypertext mechanism to be justified, these linear sequences must occasionally intersect; in other words, there must be decision points. This condition raises the problem of the meaning of the reader's choices and of the function of links.

Function of Links In Hypertext

Generally speaking, the interactivity of hypertext is external rather than internal (as defined in chapter 7), since the user usually does not act as a character, and exploratory rather than ontological, since the user does not determine the destiny of the storyworld but explores possibilities already encoded in the text.[6] Level of application is the second of the levels defined in chapter 7. Clicking on links may fulfill a variety of functions, some restricted to literary/narrative forms, others also found in informational texts. Here is a tentative list of these functions.

- Propose, and let the reader explore, many possible futures for the textual world. This use of hyperlinks asks the reader to adopt the prospective orientation of simulation. Each of the forking paths determined by the links corresponds to a possible world in the alethic/temporal sense—that which could still happen, given the present situation and the laws of the textual world. This kind of interactivity, which determines the plot, is typical of *Choose Your Own Adventure* stories.

- Propose, and let the reader explore, alternative versions of a reasonably solid core of facts. This approach interprets the notion of possible world in the epistemic sense. Insofar as each path offers to the reader a particular point of view on a common actual world, it represents the "belief-world" of a distinct individual.
- Shift perspective on the textual world. This can mean that the reader might open a new window, follow another plot line, or switch focus to a different character and enter a new private world.
- Control the reader's progress in the discovery of facts. Even in a nonlinear text, it may be necessary to read certain fragments before others make sense. By creating systems of links, authors establish complex systems of dependencies between lexia. "Guard fields," "multi-headed," and "multi-tailed" links may, for instance, ensure that the reader has followed a certain path before reaching a certain node.[7]
- Suggest analogical relations between segments. One of the interpretive strategies most characteristic of hypertext fiction (as opposed to print novels) is to ask why the author placed a hyperlink between the preceding and the current lexia. Answering this question often necessitates a demanding act of memory, since the reader must compare a present with an absent lexia. But this difficulty can be alleviated by split screens that show both of the lexias.
- Allow the user to blow up certain screens or passages and get a closer look. This is common practice in graphics software, but the idea can also be used with text and multimedia works. In a narrative, the system could offer a choice between a "scene" and "summary" representation of an episode or enable the reader to skip or expand descriptions.
- Interrupt the flow of narration, disrupt, frustrate, puzzle, undermine certainty, subvert or mock the text on the other end of the link, withdraw facts, and place fictional worlds "under erasure."[8]
- Provide background information, explanations, supporting material, and intertextual references.
- Keep the textual machine going. When the reader's choices are fully blind, the purpose of clicking on one link rather than another is simply to get more text to the screen. In this case, interactivity is nothing more than

> a user-operated random number generator. In the electronic poetry of Jim Rosenberg or John Cayley, for instance, the reader animates certain words or causes certain sentences to appear by simply moving the cursor around the screen and randomly clicking. The same principle can be used to animate images; as the cursor moves over certain areas of the picture, it triggers unforeseeable metamorphoses.

Depending on the particular purpose, the links will be opaque or involve navigational aides (such as meaningful highlighted keywords). Access to a given link will be either context free or made conditional on the visiting of other links. Links may be visible or hidden like Easter eggs inside text and pictures. And finally, some links are permanent, while others self-destruct after a visit.

Conceptualizations of Hypertext Interactivity

To develop a sense of involvement with a computer program, the user needs a scenario that casts him into a role and projects his actions as the performance of concrete, familiar tasks: writing, editing, drawing, sending mail, building cities, or killing dragons to save a princess. In hypertext, this conceptualization is problematic. Clicking on links is not really reading because reading is something that is done with the eyes and the mind and not with the hand. Nor is clicking a matter of turning the pages because this analogy would revert to the mode of reading characteristic of the codex book. Yet the need for interpretive scenarios that give meaning to the reader's engagement with the text is particularly urgent if readers are to accept a form of textuality as different from standard print forms as hypertext.

One of the first theorists of electronic media to recognize the importance of metaphorical mappings for the psychology of computing was Brenda Laurel. In her pioneering book on the strategies of interface design, *Computers as Theatre,* she developed the theater metaphor in the hope of replacing the traditional and rather uninspiring perception of the computer as a tool with "the idea of computer as a representer of a virtual world or system, in which the person may interact more or less directly with the representation" (127). When the metaphor is adapted to the case of interactive textuality, it casts the screen as a stage, the run of the text as performance, and the user as the director who summons actors to the stage and releases their lines through the click of the mouse.

Another popular interpretive scenario is a development of the spatial metaphor. In this mapping, the text as a whole is a territory, the links are roads, the textual units are destinations, the reader is a traveler or navigator, clicking is a mode of transportation, and the itinerary selected by the traveler is a "story." Since every reader follows a different itinerary, every reading session "writes" a different narrative, and through his agency the reader becomes the "author" of his own adventures. The spatial metaphor supports different scenarios depending on whether textual space is conceived as a "smooth" expanse that the reader cruises for the pleasure of the trip or as a "striated" space of freeways whose sole purpose is to take the traveler to a destination.[9] (Here again I refer to Deleuze and Guattari's famous dichotomy.) In a smooth space environment, the reader is driven by an obsession to get further, either fortified or dampened in this drive by the thought that his desire to exhaust all the links cannot be satisfied. In a striated space, the reader gives herself a goal, such as reaching the center of a labyrinth or finding the exit, and her relation to the text is very much that of a player who hopes to beat a computer game.

The reader's role in hypertext can also be conceived along the lines of the supermarket-shopping experience. In this scenario, suggested to me by Christopher Keep ("The Disturbing Liveliness of Machines," 175), the reader browses along the links, takes a quick look at the commodities displayed on the screens, and either drops them in his shopping basket for careful study or moves on to other screens. This reader does not feel compelled to read the text in its entirety nor to pay attention to every screen because he sees the text not as a work held together by a global design but as a display of resources from which he can freely pick and choose. I was tempted to take this attitude with hypertexts such as Stuart Moulthrop's *Hegirascope* and Mark Amerika's *Grammatron.* Both texts gave me the impression of a loose collection of drafts and writing samples whose appeal resides in the imaginative quality, wordplay, or visual effects of the individual segments rather than in a sustained structure of expectations. In such an approach, the links between segments are not regarded as a dimension of meaning but as a mode of transportation to the next screen to be sampled. With its modular structure (small stories contained in individual lexias), Deena Larsen's *Marble Springs* also lends itself to this type of reading. In a variant of the supermarket scenario, the reader puts lexias in his shopping basket not to consume them individually but to use them as material to construct his own stories. This is how George Landow envisions the reader's ability to make narrative sense out of a nonlinear collection of lexias: "In a hypertext environment a lack of linearity does not destroy narrative. In fact, since readers always, but particularly in this environment, fabricate their own structures, sequences, and meanings, they have surprisingly little trouble reading a story or reading for a story" (*Hypertext 2.0,* 197). In this approach, the reader imagines a linear order that overwrites the system of links.

A third conceptualization relies on the metaphor of the kaleidoscope. In this scenario, the text consists of a collection of fragments that can be combined into ever changing configurations through the random choices of the reader. A variant of this analogy is the image of the construction kit, which attributes purposeful agency to the user and thus leads into the "wreader" theories mentioned earlier. In a kaleidoscope, the built-in mirrors create a symmetry that guarantees the aesthetic appeal of all permutations. By analogy, the measure of success for hypertext authors would be to design a system in which, to quote Bolter again, "every path defines an equally convincing and appropriate reading" (25). The metaphor of the kaleidoscope presupposes that sequence is significant, otherwise the system would not be able to produce ever new images. Hypertext, we must remember, is not literally spatial; all segments must be read in succession. The significance of sequence is easily demonstrated in a narrative framework: "Mary had a baby," "Mary married Joseph," "Mary lost her virginity" tells an entirely different story than the same propositions read in the reverse order. (But we must supply many inferences to make it meaningful, especially when we read it in the sequence listed above!) Michael Joyce implicitly endorses the metaphor of hypertext as kaleidoscopic storytelling machine when he writes (with a probably broader concept of story in mind than the one illustrated above): "Reordering requires a new text; every reading thus becomes a new text. . . . Hypertext narratives become virtual storytellers" (*Of Two Minds,* 193). When the system of links is reasonably well developed, the author cannot predict the paths that will be taken beyond the transitions out of or into a given segment. This prompts Gunnar Liestøl to write, "In reading hypertext fiction the reader not only recreates narratives but creates and invents new ones not even conceived of by the primary author" ("Wittgenstein, Genette," 98). Another implication of the kaleidoscope metaphor is that the reader *cannot* freely pick and choose; every new story consists of the same elements arranged in a different configuration. To see a story (or, at least, a meaningful juxtaposition) in whatever comes in her way, the reader must be a priori willing to supply whatever missing information is needed to give meaning and coherence to the sequence of lexias created by his decisions. For every trip into the text to generate a genuinely new and autonomous story, the reader must erase from memory the knowledge gathered and the expectations created from previous reading sessions.

The kaleidoscope model works better with poetic texts in which the meaning of the sequence is not narrative but lyrical, that is, not logical, causal, and temporal but associative, thematic, and quite tolerant of incongruous juxtapositions. In Raymond Queneau's print text *Cent mille milliards de poèmes,* for instance, 10^{14} different sonnets can be generated by combining the lines of ten different sonnets of fourteen lines each. Here the notion of coherence is much more flexible than in narrative, and it is as easy to see meaning in each sequence of lines as it is to see shapes in the blots of a Rorschach test. But consider what would happen if the various combinations of the kaleidoscopic text were interpreted along strict narrative lines—*y* read after *x* meaning that *y* followed *x* in time. In certain traversals of Stuart Moulthrop's *Victory Garden,* the reader first learns that a character, Emily Runbird, has been killed by a missile in the Gulf War; then the reader encounters passages that tell of Emily's college days and of her love affair with a professor. If the sequence had temporal meaning, this would imply that Emily comes back from the dead, an interpretation that leads to all sorts of unwanted consequences, for *Victory Garden* is on the whole a realistic narrative, not a fantastic tale nor a religious myth. No reader will come to the absurd conclusion that Emily has been resurrected; the most elementary narrative competence tells us to read the passages narrating Emily's college years as a flashback compared to those that presuppose her death. This strategy may seem obvious, but it has two important consequences for the phenomenology of hypertext reading. First, it means that sequence can be overruled by considerations of global coherence in the construction of meaning. No matter in what order the reader encounters segments, she will assume that Emily first went to college and then died in the Gulf War. And second, if the narrative meaning of sequentiality is neutralized, this means that hypertext does not tell a different story with each reading session but, rather, creates different modes of presentation for the same underlying plot. For those who subscribe to the classical narratological distinction between story or *fabula* (what happened in the storyworld) and discourse, or *sjuzhet* (the Russian formalists' term for the verbal realization and dynamic disclosure of the *fabula*), what changes from reading to reading in the case of *Victory Garden* is not the *fabula* itself but the *sjuzhet.* The vaunted self-transforming quality of hypertext affects the reading experience much more than the textual world. Moreover, if we assume that hypertext projects a single *fabula,* rather than a radically new story in each reading session, this means that reading is cumulative and that the construction of the *fabula* can span many sessions.

The interpretive scheme that best expresses the cumulative aspect of reading is the metaphor of the jigsaw puzzle. In this scenario, the various lexias are the fragments of an exploded image that the reader tries to put together in a sustained effort. The system of links does not function as a carrier of semantic information but as one of these unnecessary obstacles that game-masters put in the way of the player. When putting a jigsaw puzzle together, the player often needs to take several looks at a piece before finding its proper place in the pictorial space. Similarly, the reader of hypertext may need several visits to the same lexia before she can gather sufficient information to place it in a meaningful environment. In the kaleidoscope metaphor, context is created by linear sequence, and every itinerary puts the visited lexia in a different context. This leads to what Liestøl has called a "constant recontextualization" ("Wittgenstein, Genette," 117). In the jigsaw puzzle metaphor, by contrast, the order in which the player looks at the pieces of the puzzle has no effect on where he puts them in the global picture. This means that the context of a lexia is formed by those other lexias that shed light upon it, quite independently of whether or not they are electronically connected. The purpose of the act of reading is not to gain an overview of the map of links but to reconstruct a network of thematic, causal, or temporal relations that bears no necessary isotopy to the system of physical connections. The difference between this scheme and the narrative version of the shopping basket is that the reader assumes that there is a specific image to be unscrambled, rather than a collection of recombinant fragments.[10]

Yet another way to deal with the fragmentation and occasional inconsistency of hypertext is the space-travel, possible worlds approach.[11] Every lexia is regarded as a representation of a different possible world and every jump to a new lexia as a recentering to another world. If we apply this strategy to Joyce's *afternoon*, there will be one world in which Peter's ex-wife and son are killed and one in which they are not, one world in which Peter causes the accident and one world in which he merely witnesses it. Each of these propositions expresses a solid fictional truth in its own reference world. If the reader takes this approach, she will not try to reconstitute a comprehensive world image, and the co-referentiality of the lexia (i.e., the fact that they are about the same individuals) will be explained through a semantic model that allows individuals to have counterparts in different possible worlds. This approach is a way to rationalize texts that present a high degree of internal contradiction, since there is no logical problem with p being the case in one world and $-p$ in another, but if every lexia is regarded as describing a self-sufficient monad, then narrative development cannot span several lexias, and the only stories that can be told are those that are contained within a node. It seems therefore improbable that readers will apply this model in a systematic fashion. Rather, they are likely to assume that certain lexia belong to many possible worlds, so that the world where Peter's wife and son are dead and the world where they are alive will share a common history built on the basis of lexias that are compatible with both states of affairs, such as those that assert the existence of an ex-wife and son for Peter.

Which one of these scenarios will be preferred depends as much on the reader's individual disposition as on the nature of the text. Easily detectable relations between adjacent segments suggest the metaphors of travel or of the kaleidoscope because they give meaning to sequence, links that break up semantic continuity tend to redirect the reader's attention toward a global coherence to be reconstituted in jigsaw puzzle fashion, and relatively autonomous lexias will promote the shopping-cart scheme. The supermarket metaphor is also best suited for readers with a short attention span. Meanwhile, the kaleidoscope metaphor should appeal to those who are fascinated with combinatorics, while the jigsaw puzzle metaphor will be favored by readers naturally inclined to approach literary texts as problems to be solved. The structural fluidity of most hypertexts is usually compatible with several conceptual schemes, but this fluidity also prevents the reader from feeling totally secure in any one of them.

One common feature of all the interpretive scenarios described above is that they present the reader's conceptualization as an attempt to overcome the text's fragmented appearance and to restore some kind of coherence—either that of a spatial landscape or of a temporally and causally ordered narrative. Some might object that such strategies try to correct and therefore implicitly reject the fundamentally disjointed, fragmented, and untotalizable "nature" of hypertext. The basic dilemma, as I see it, is whether memory and something worth calling "understanding" can live under such radical conditions of segmentation or whether linearization and totalization (even if this means of only a subset of the text) are elementary sense-making activities.

Hypertext and Immersion

In this section I discuss the immersive potential of hypertext with respect to temporal, spatial, and emotional immersion. I return to these types in chapter 10 when I evaluate internal and ontological forms of interactivity.

Temporal Immersion

This is the most problematic of the three types of immersion. The temporal effects described by Meir Sternberg—suspense, curiosity, and surprise—depend on a precise management of the reader's knowledge. For suspense, the reader should be able to anticipate at least two possible developments on the basis of past events; for surprise, past events should make one future development look inevitable, but the story should take an alternative, unpredicted path; and for curiosity, the reader should know the outcome, but she should not know through what path the story got there. By permitting readers to reach certain nodes through various routes and to acquire different information on the way, the network structure of hypertext prevents a strict control of the reader's knowledge. I must, however, confess to experiencing some suspense in Michael Joyce's *afternoon* when the narrator makes several attempts to find out if the victims of an accident he has witnessed are his ex-wife and son. This suspense is made possible by having the reader visit the node that relates the witnessing of the accident before she reads about the narrator's frantic phone calls to the hospital and to various people, but the same effect could have been achieved more easily in a classical linear organization. To create some suspense, Joyce had to work against rather than with the properties of the medium, since he had to limit the reader's choices for at least one area of the underlying network. A similar strategy could have been used to resolve the suspense, but the author chose not to do so: the reader never learns the truth about the accident.

Spatial Immersion

The spatiality of hypertext can be conceived in two different ways: as a feature of form and as a feature of content. On the level of form, spatiality resides in the map of the text, this is to say, in the configuration of the database out of which the reader pulls information by activating certain links. This kind of spatiality is unique to hypertext. On the level of content, by contrast, spatiality is simply the space of the world evoked by the text, if indeed the textual information coheres into a world, and it is a feature shared by all representational texts. It is only on the level of the storyworld that the reader's experience can be called immersive in a reasonably literal way (keeping in mind that immersion is itself a metaphor), but the reader's movements through the hypertextual network can be given a figural meaning that captures a postmodern experience of space.

For Fredric Jameson, postmodern space is an alienating, self-transforming expanse that offers neither rest for the body, refuge for the soul, nor landmarks for the mind: "[This] latest mutation in space—postmodern hyperspace—has finally succeeded in transcending the capacities of the individual human body to locate itself, to organize its immediate surrounding perceptually, and cognitively to map its position in a mappable external world" (*Postmodernism,* 44). This postmodern space inspires a feeling of being lost and of aimless wandering. Both of these experiences are reflected in the reader's blind progression through the labyrinthine structure of hypertext. For Marc Augé, the characteristic space of "supermodernity," as he calls the present time, is a nonplace that we traverse on our way to somewhere else and that we come to inhabit, since the postmodern condition is a state of perpetual transit. These nonplaces are called airport, subway, freeway, and the network of information superhighways that crisscross cyberspace, the most nonplace of them all.[12] Cybernauts and hypertext readers spend most of their time clicking on the nonplaces of the links, never dwelling for long on a textual segment because each of these segments is less a destination than a point of departure for other, equally elusive destinations.

The hypertextual network may convey an exhilarating sense of mobility, but the cost of embracing space in its globality is an alienation from its locality that prevents growing roots in any given site. Links are a constant temptation to move beyond the present screen. As Michael Heim observes, "Hypertext thinking may indeed reveal something about us that is agitated, panicky, or even pathological. As the mind jumps, the psyche gets jumpy or hyper" (*Metaphysics,* 40). One consequence of the segmented structure of hypertext is that the lexias are rarely long enough to let an atmosphere sink in. Every time the reader is called upon to make a decision, she must detach herself from the narrative "here and now" and adopt a point of view from which she can contemplate several alternatives. Once the choice is made, the reader may regret her decision and be haunted with the "could have been." What Gareth Rees writes of his experience of tree fiction is even more to the point in the case of a more complex network: "I think that as readers we are not ready for tree fiction: I know that when I read such a story, I want to find out all the consequences of every decision, to read everything that the author wrote, fearing that all the interesting developments are going on in another branch of the story that I didn't investigate. I want to organize the whole story in my mind" ("Tree Fiction"). The body of the reader's imaginary persona in the fictional world would have to undergo a dismembering to take all the roads at the same time and to overcome the nagging feeling of missing something along the way. If spatial immersion is an intimate relation of a body to a nurturing environment, it cannot be experienced by a *corps morcelé,* as Christopher Keep describes the hypertextual reconfiguration of the body.

Emotional Immersion

Once again the question is not whether the reader of hypertext can develop the kind of affective relations that lead to feelings of happiness or sadness when things turn out for the better or for the worse for a certain character but whether or not interactive mechanisms can be used to enhance this emotional participation. Janet Murray writes that the sense of the irrevocability of life is alien to the spirit of interactive art: "Our fixation on electronic games and stories is in part an enactment of [a] denial of death. They offer us a chance to erase memory, to start over, to replay an event and try for a different resolution. In this respect, electronic media have the advantage of enacting a deeply comic vision of life, a vision of retrievable mistakes and open options" (*Hamlet,* 175).[13] One of the trademarks of the spirit of comedy is a playful detachment from the characters that precludes an affective investment in their fate. This detachment is strengthened by the knowledge that the character's life is simultaneously acted out in several possible worlds and that if we do not like one of these worlds we can always jump to another. Emotional immersion requires a sense of the inexorable character of fate, of the finality of every event in the character's life, but as Umberto Eco observed in a radio interview, this outlook is fundamentally incompatible with the multiple threads generated by interactive freedom. "A hypertext can never be satisfying," Murray quotes Eco as saying, "because 'the charm of a text is that it forces you to face destiny'" (296). Julian Barnes, in *Flaubert's Parrot,* has a solution for this problem:

> In life, we make a decision—or a decision makes us—and we go one way; had we made a different decision . . . we would have been elsewhere. The novel with two endings doesn't reproduce this reality: it merely takes us down two diverging paths. It's a form of cubism, I suppose. And that's all right; but let's not deceive ourselves about the artifice involved.
>
> After all, if novelists truly wanted to simulate the delta of life's possibilities, this is what they'd do. At the back of the book would be a series of sealed envelopes in various colors. Each would be clearly marked on the outside: Traditional Happy Ending; Traditional Unhappy Ending; Traditional Half-and-Half Ending; Deus ex Machina; Modernist Arbitrary Ending . . . and so on. You would be allowed only one, and would have to destroy the envelopes you didn't select. That's what I call offering the reader a choice of endings. (89)

Yet it would be simplistic to deny hypertext any power of emotional appeal. I was for instance deeply moved by the multimodal hypertext *Juvenate,* which explores the inner life of a dying man haunted by the sounds and images of a world from which he is now physically isolated in his hospital bed (Ryan, *Avatars*), as well as by the anguished attempts of the narrator of *afternoon* to find out about the fate of his ex-wife and son. Rather than using the network of links and nodes to suggest the consequences of different decisions, these texts exploit the forking paths of the system to represent the stream of consciousness of a character whose thoughts return obsessively to certain scenes or objects. Entering a character's consciousness creates an identification with the character which can lead to a powerfully emotional immersion. Whether or not the hypertextual mechanism allows or even favors emotional immersion cannot therefore be answered in a general way; it depends too much on the subject matter and on the author's use of the mechanism.

Digital literature relies heavily on the software tools available to authors. The heyday of hypertext fiction corresponded to the development and adoption, in the 1990s, of Eastgate's Storyspace program. In the 2000s, the increased multimodal capabilities of computers and the introduction of Flash and Director (especially Flash) spelled the end of what Robert Coover called "the golden age" of electronic literature: the only purely verbal texts in the two collections published by the Electronic Literature Organization are those that predate 2000. The shorter texts of Flash are even less conducive to narrativity and world-creation than the lengthy texts of Storyspace; this explains in part why worldless texts that flirt with various forms of dysfunctionality (such as those described in chapter 6) are much more widely represented in the two ELO collections than genuinely narrative texts. Nowadays the hegemony of Flash is breaking up, and digital authors are turning toward other supporting technologies: mobile phones, touchscreen tablets, geopositioning systems, and augmented reality, each of which has the potential of producing narrative applications. Meanwhile, commercial computer games, able to rely on the most advanced technology in the visual, kinetic, and artificial intelligence domains, are steadily perfecting their storytelling abilities, but all the technological resources in the world do not help solve what I call in the preceding chapter the interactive paradox because it is an essentially logical and creative issue. In chapter 10, I outline a poetics of interactive narrative that offers no definitive solutions to the interactive paradox, for paradoxes are by definition insolvable, but at least proposes acceptable compromises between narrative design and user freedom. These compromises rely on much more immersive forms of user participation than the purely external-exploratory interactivity of hypertext fiction. First, however, I explore examples of this combination in nondigital environments.

IV: Reconciling Immersion and Interactivity

NINE

Participatory Interactivity from Life Situations to Drama

> I . . . remember vividly to this day the terrible shock of such a recall to actuality [from within a fictional world]: as a young child I saw Maude Adams in Peter Pan. It was my first visit to the theater, and the illusion was absolute and overwhelming, like something supernatural. At the highest point of the action (Tinkerbell had drunk Peter's poisoned medicine to save him from doing so, and was dying) Peter turned to the spectators and asked them to attest their belief in fairies. Instantly the illusion was gone; there were hundreds of children, sitting in rows, clapping and even calling, while Miss Adams, dressed up as Peter Pan, spoke to us like a teacher coaching us into a play in which she herself was taking the title role. I did not understand, of course, what had happened; but an acute misery obliterated the rest of the scene, and was not entirely dispelled until the curtain rose on a new set.—*Susanne K. Langer*

The interactivity of the texts widely regarded as "electronic literature" tends to develop a self-reflexive dimension that alienates the reader from the storyworld or, in the most extreme cases, attracts so much attention to the signifiers that their ability to summon a world to the imagination is altogether blocked. For interactivity to be reconciled with immersion, it must be stripped of any self-reflexive dimension. To find out how to avoid the "holographic effect" that forces us to choose between playing with signs and plunging into the depth of worlds, let us take another look at the ideal VR experience. Why is it that in VR, far from destroying illusion, the possibility of interacting with the virtual world should reinforce the sense of its presence?[1] Let us first note that in real life (RL) also, freedom to act enhances our bond to the environment. The main difference between VR and RL, on one hand, and textual environments, on the other, is the semiotic nature of interactivity. In a textual environment the user deals with signs, both as tools (the words or icons to click on) and as the target of the action (the text brought to the screen as the result of clicking), but in RL and VR all action passes through the body. Of course, it takes a hand to click or write in textual environments, and the entities encountered in VR are ultimately digital signs, but, if realized, the ideal of the disappearance of the medium means that the VR user experiences at least the sensation of a direct encounter with reality. The hand that turns the pages of a book or that clicks on hypertext links does not belong to the textual world, but the body that moves around in a VR installation writes, or rather performs, the "history" of the virtual world. We cannot say that "*x* clicks on link *p* at time *t*" is an event in any of the worlds of Michael Joyce's *afternoon*—though it is an event in the reader's reading, which takes place in the real world—but in a VR installation, "*x* turns his head to the left," "*y* grabs an object," or "*z* takes an action that steers the plane toward the Earth" are history-making events. VR and RL thus offer a mode of action in which the body can be much more directly and fully involved with the surrounding world than through conscious symbolic manipulation.

Moreover, as we have seen in chapter 2, the worlds of VR are experienced as existing autonomously because they can be explored through many senses, particularly through the sense of touch. As the story of St. Thomas demonstrates, tactile sensations are second to none in establishing a sense of the alterity that convinces the mind of the objective existence of reality. The perceptions of the various senses are not isolated in the discrete windows of a hypermedia display, as they are in hypertext and World Wide Web pages, but fused in a global experience that enables the user to apprehend the virtual world through many facets at the same time.

The VR display's fluidity is another critical factor of immersivity, for the world does not offer itself to the senses in broken-up packets of information. In a textual environment interactivity is made possible by the chopping down of data into the discrete fragments of "lexias" or "textrons"; in a VR installation, by contrast, the packets of data that respond to the user's actions are so small that they give the impression of a continuous evolution.

Finally, the corporeal participation of the user in VR can be termed world-creative in the same sense that performing actions in the real world can be said to create reality. As a purely mental event, textual creation can be called a creation ex nihilo in the sense that it excludes the creator from the creation: authors do not belong to the world of their fictions.[2] But while a mind may conceive a world from the outside, a body always experiences it from the inside. As a relation involving the body, VR's interactivity immerses the user in a world experienced as already in place; as a process involving the mind, it turns the user's sojourn in the virtual world into a creative membership. For an agent embodied in a multidimensional environment, exploratory and ontological interactivity can no longer be rigidly distinguished because being-in-a-world is a way to make this world flow out of the acting body, a way to create it.

By arguing that the key to immersive interactivity resides in the participation of the body in an art-world, I do not wish to suggest that interaction should be reduced to physical gestures but rather that language itself should become a gesture. As is the case in dramatic performance, the participant's verbal contribution will count as the actions and speech acts of an embodied member of the fictional world. Rather than performing a creation through a diegetic (i.e., descriptive) use of language, these contributions will create the fictional world from within in a dialogic and live interaction with its objects and its other members. Whereas merely immersive art is a representation of a fictional world, the reconciliation of immersion and interactivity will propose a genuine *simulation.*

Where can we turn, besides VR, for models of an enriching experience of immersive interactivity? Long before electronic technology made it possible to "[take] your body with you into worlds of imagination," to quote once again Brenda Laurel's seductive characterization of the power of VR ("Art and Activism," 14), these worlds were made accessible to either flesh and blood or virtual bodies through a variety of art forms and cultural activities.

Children's Games of Make-Believe

Games of make-believe provide a compromise between the "game" and the "world" aesthetics, since their purpose is to create a world in which to play. This purpose makes them uniquely qualified to reconcile immersion and interactivity. My strongest memory of being corporeally and creatively involved in a fictional world is a game that we used to play in kindergarten, "Who's Afraid of the Big Bad Wolf?" One player was the wolf and the others were pigs. A delimited territory, such as a manhole cover, was the house of the wolf; all around were the woods. At the beginning of the game, the wolf was sleeping in his house. The pigs would gather near his door and taunt him with a chant: *Promenons-nous / Dans les bois / Quand le loup / N'y est pas. / Loup y es-tu? / Que fais-tu?* (Let's walk in the woods when the wolf is not there. Wolf, are you there? What are you doing?) The wolf would respond to each repetition of the chant with an improvised line: "I am getting out of bed, putting on my socks, brushing my teeth, adjusting my toupee," and so on. The responses had to follow the script of getting up and leaving the house for work, but the player was free to add as many colorful details as he wanted, both for the sake of their inherent creativity and as retarding devices. The pigs would react with delighted screams of terror or with renewed taunting. The suspense rose as the wolf got closer and closer to being ready. Finally, he would roar some terrifying line, storm out of the house, and go after the pigs as they scattered into the woods. The pig that got eaten became the next wolf.

In this game, both the wolf and the pigs had the freedom to improvise, and the creativity of their contributions was an element of pleasure for all players. At the same time, however, the improvisation was monitored by a narrative scenario with a proper dramatic development: exposition (the wolf getting up), complication (he is taunted), climax (he storms out of his house), and resolution (he takes revenge on one of the pigs while the others get away). What made the experience so immersively pleasurable is that our creative interaction started not in a vacuum but in a designed and rule-governed environment that offered, as should any efficient "prop in a game of make-believe," a potent stimulant to the imagination.

The Fair and the Amusement Park

If there is an environment in which you can literally take your body with you into worlds of the imagination, it is the amusement park. We may dismiss this culturally invasive avatar of the fair as an attempt to recreate the communal street life of earlier days in a fenced-off and fully commercialized space of three-dimensional postcards, but with its multiple offerings, diversified geography, parades and masquerades, and mosaic of sensory stimuli, it also stands as a testament to the postmodern fascination with the playful spirit and protean nature of the carnivalesque. Jay Bolter and Richard Grusin capture the immersive dimension of theme parks through these observations: "The parks could appeal to the immediacy of physical presence. The rides themselves offer sound, light, and tactile sensations, as amusement parks have always done, and the themes and narratives associated with the rides, the animated characters that roam the park, and the themed architecture all give the young visitors the exhilaration of being physically surrounded by the media" (*Remediation*, 172).

Glorianna Davenport's eloquent description of the affinities of Coney Island, the ancestor of all modern-day fairs, with the ideals of electronic entertainment places more emphasis on the interactive aspect of the amusement park experience. For Davenport, Coney Island is a model of immersive interactivity because once the visitor has stepped over the boundaries of its magic world, she feels instantly at home, and because this world goes out of its way to show her what to do in it.

Through this internal guidance, the world of the fair displays choices of action without breaking the illusion.

> Like the World Wide Web of today, Coney Island was an anarchical sprawl, lacking any form of urban planning or overarching design. Browsing, "grazing," and the delight of accidental encounter were the rule rather than the exception. It offered something for every taste: the physical thrill of high G-forces as a roller coaster hurled your body along a wildly twisting path; the sexual titillation of exotically costumed showgirls shimmying on stage; the long-term, episodic drama of premature babies struggling for survival in Martin Courtney's "Child Hatchery" (which introduced and popularized the use of incubators in hospitals).
>
> Most notably, Coney Island was exceedingly "user friendly": Visitors needed no tutorial sessions or user manuals to enjoy its attractions. In the distance, the potent sight of enormous rides and ornate buildings lured potential customers. Just as importantly, these huge and uniquely delineated structures defined a sort of cognitive map—or a preliminary shopping list, at least—to assist navigation on the ground. Close-up, Coney Island's pleasures became more tangible and human-scale. Colorfully dressed barkers and shills lined the streets, their sole purpose to steer passersby into a particular attraction. Once inside, rich detail, meaningful travel, and the delight of discovery marked every moment of your experience.
>
> Coney Island provides an outstanding model of "virtual reality," a successful symbiosis of human entertainers and constructed mechanisms dedicated to human pleasure. How many hours have you spent in front of your personal computer wishing you were at Coney Island instead? ("Care and Feeding," 8)

Baroque Art and Architecture

It is almost a truism to say that architecture is an art that inscribes the body in its world. Its material is space, and its purpose is to arrange this material for the recipient to inhabit, even if the sojourn is only a temporary visit. It is no less obvious that architecture is an art that requires a physically active involvement. The building is designed for a living body, and the body must perform a walk-through and walk-around to experience the spatial design under all the points of view that matter to the intended recipient (typically those points of view within reach of a normal-sized adult). But the immersive power of architecture is usually tempered by abstract shapes that do not engage the imagination. This is why Baroque churches, with their predilection for the curves of seashells and violins, their extravagant decoration, their exuberant use of organic forms, and their trompe l'oeil effects, offer a better prefiguration of the VR experience than the architecture of any other age.[3] My most memorable sensation of interactive immersion—if we interpret *interactive* in a metaphorical way, as a mind-set rather than as physical intervention—came about during a visit to the Rococo-style abbey church of Zwiefalten, located in the Swabian Jura halfway between Munich and the Rhine.

The church at Zwiefalten embodies the realm of the fake and the triumph of appearance. The gold of the crown of Mary is made up of only a thin layer of precious metal, the "marble" of the columns is painted plaster, and many of the architectural details are themselves painted on the walls. The Baroque age is notorious for its suspicion of mere appearances—in stark contrast to present times—but it also understood the value of the fake as a "prop," not in a game of make-believe (to parody Kendall Walton) but in the business of creating belief. A church, by definition, is the site of a lived interaction between man and God. In the Baroque age this interaction was conceived and staged as an experience that involved the whole of the person, this whole that Saint Ignatius of Loyola described as an indivisible "compound of body and soul" (*Spiritual Exercises*, 136). In his *Spiritual Exercises*, Ignatius advocated a program for developing and strengthening faith that used the five senses of the body as stepping stones toward the involvement of what he considered the two senses of the soul: the will and the intellect. The "exercitant" was asked to imagine the biblical texts, such as the story of the Nativity, by building step by step (Ignatius was obsessed with subdivision) a mental simulation that began with visualization, added hearing, smell, and touch (taste was harder to include), and ended with a sense of corporeal presence on the narrated scene which provided an intense emotional participation in the sacred events.

The religious art inspired by the Counterreformation offers the concrete expression of the spirituality of Ignatius. The brightly colored interiors of Baroque churches, their illusionist paintings, tonitruous organ music, burning incense, alabaster statues that invite caressing, and even relics (or their containers) occasionally offered to the touch speak to all of the senses that Ignatius wanted to involve in the religious experience. Through its insistence on the corporeal—there are bodies crammed into every niche and jamming every picture—the style proposed a literalization of the doctrine of the Word made flesh.

There is no need to dwell here on the well-known sensuality of Baroque art, on its almost sadistic depiction of the suffering flesh of Christ, or on its representation of the mystical union as orgasmic trance (consider the statue of St. Theresa by

Bernini in Rome's Santa Maria Della Victoria church). The Baroque age solved the Christian conflict of sensuality and spirituality by presenting the marriage of God and the soul as a deeply sexual experience. Whereas the design of Gothic churches integrates the divine as a purely spiritual light that enters the dark interior of the building (a symbol of the human soul) from above and from outside, the Baroque church surrounds the visitor with the radiant, almost physical presence of God. The light does not come from an external source, it originates inside the church itself, in the whitewashed walls, golden frames, lighted candles, and gilded moldings of the altar. Walking down the aisles, past numerous alcoves that represent in three-dimensional displays the various episodes of the Passion, the visitor wanders in a story space inhabited by lifelike and palpable incarnations of biblical characters. This dramatic organization of space, typical of all Catholic churches, fulfills Celia Pearce's recommendation that architecture be treated as a narrative art and the visit to a building be structured as a meaningful succession of events (*Interactive Book*, 25–29).

At Zwiefalten, the sense of corporeal involvement in the religious experience is reinforced by numerous trompe l'oeil effects that expand the church's physical space and invite the visitor to ascend into virtual realms. The cupola's ceiling is occupied by a fresco painted from the perspective of an observer situated directly under the scene (known as *Di sotto in sù*), exactly where the actual spectator happens to be physically located.[4] As in all trompe l'oeil art, this painting "continues the architecture of the building itself" (Bolter and Grusin, *Remediation*, 25). The effect of the coincidence between actual and virtual observer is a powerful sense of being pulled into pictorial space. The fresco represents the stages of a tortuous ascension toward heaven, which could be said to emulate the levels of a computer game, though the inspiration runs in the opposite direction; it is computer games that emulate the stages of religious experience. The levels of heaven are connected by the flights of a gigantic staircase that presents the quest for the kingdom of God as an active process and continuous effort. At the apex of the real-world cupola, in the picture's center, a ray of lights breaks through the clouds and pierces the hearts of those who have traveled so far. From there on, believers will be literally pulled by God into the upper reaches of heaven. In this dynamic vision of life, which makes it impossible to contemplate the painting without mentally traveling the path from Earth to heaven, the grace of God touches only those who work for their own salvation.

Umberto Eco has superbly captured this power of Baroque art to pull the spectator into its own movement and to simulate a corporeal mode of participation.

> Baroque form is dynamic; it tends to an indeterminacy of effect (in its play of solid and void, light and darkness, with its curvatures, its broken surfaces, its widely diversified angles of inclination); it conveys the idea of space being progressively dilated. Its search for kinetic excitement and illusory effects leads to a situation where the plastic mass in the Baroque work of art never allows a privileged, definitive, frontal view; rather, it induces the spectator to shift his position continuously in order to see the work in constantly new aspects, as if it were in a state of perpetual transformation. (*Open Work*, 7)

Art of the fake put in the service of religious truth; art of curves, labyrinths, and morphing shapes that launch a dance of the imagination—the Baroque is VR in veneer and stucco.[5]

Ritual

The style of the Baroque church may suggest immersive interaction, but as a ritual, the ceremony performed in the building aims at accomplishing a much more complete and literal fusion of the two modes of binding to a world. In its religious form, ritual is a technique of immersion in a sacred reality that uses gestures, performative speech, and the manipulation of symbolic objects—symbolic at least until the ceremonial protocol succeeds in establishing direct communication between the human and its Other. As Mircea Eliade, the great historian of religions, has convincingly argued, religious ritual is the reenactment of the foundational events that are commemorated in myth. Through the exact repetition of what the gods did in the beginning (*in illo tempore*), the community is transported into mythical time and undergoes spiritual rebirth: "Ritual abolishes profane, chronological Time and recovers the sacred Time of myth. Man becomes contemporary with the exploits that the Gods performed *in illo tempore.* On the one hand, this revolt against the irreversibility of Time may help man to 'construct reality'; on the other, it frees him from the weight of the dead Time, assures him that he is able to abolish the past, to begin his life anew, and to re-create his world" (*Myths*, 139).

As they reenter primordial time, the participants in the ritual experience the live presence of the gods as an infusion of creative power. In this state of interactive immersion in the sacred, they attain a status that can be described as coauthorship of the cosmos. From a purely semiotic point of view, the climax of ritual is an event of transubstantiation, by virtue of which symbols are metamorphosed into what they represent. Through this transubstantiation the performing bodies of the participants become the incarnation, not just the image, of mythical beings, and the commemorative language of absence is replaced by the jubilation of live presence. In the Catholic mass, the passage from symbolic re-presentation to literal enactment occurs during the moment known as consecration, when the bread and wine of the communion are miraculously transformed into the flesh and blood of Christ. The participation of the community in the sacred, which rewards it for the exact performance of the prescribed actions, thus prefigures the "postsymbolic" communication that forms the lofty ideal of the most exalted prophets of VR technology.

The affinities between VR and ritual are indeed a common theme in the discourse devoted to the medium. The comparison seems to grow spontaneously out of the atmosphere of religious fervor that has surrounded the early development of cyberculture. But whereas the above discussion conceives ritual as an action that produces immersion through transubstantiation, most of the writers who have explored the analogy invoke the archetypal scenario of the rite of passage and initiatory journey. For Howard Rheingold, the first virtual realities were the painted caves of the Paleolithic age, the best known of which is the cave of Lascaux. Though Rheingold's book *Virtual Reality* appeared too early (1991) to establish connections with this newer technological development, the prehistoric cave prefigures the form of VR in which images are projected onto the walls of a small room, rather than being made visible and palpable through goggles and data gloves. In reference to Plato, the VR room is known as the CAVE, an acronym for Cave Automatic Virtual Environment (Heim, *Virtual Realism*, 26).

Building on the theories of the paleontologist John Pfeiffer, Rheingold imagines a ceremony in which carefully selected novices were led into the caves by a group of shamans to learn the secrets of toolmaking, the most important breakthrough in the history of the human race. Rheingold speculates that when the candidates entered the underworld of the cave, after hours of crawling through dark shafts, the figures on the walls illuminated by the flickering light of torches took on magical life and induced an altered state of consciousness, thereby putting the candidates in the proper state of mind to be initiated into technological secrets (*Virtual Reality*, 379ff.). This scenario includes the three standard moments of the rite of passage: (1) separation from the profane world; (2) journey in a dangerous place, site of a symbolic death; and (3) spiritual rebirth and return to the everyday world. The pivotal moment in this protocol is the immersive experience that cleanses the mind and makes room for the acquisition of new knowledge.

For Brenda Laurel, the ceremonial model for the spiritual potential of VR is the cult of Dionysus, as well as the kivas of the Anasazi culture of the Four Corners area of the southwestern United States. In an oral response to Rheingold she says, "The transmission of values and cultural information is one face of VR. The other face is the creation of a Dionysian experience. The piece I find important to both of these functions is this notion of being in the living presence of something. With the ceremony of the kiva, one is in [the] living presence not just of other people but of an event that is happening in real time" (quoted in Rheingold, *Virtual Reality,* 385). The presence of the gods can be compared to the telepresence of VR because it breaks the boundary between the realm of the human, located *here,* and the realm of the divine, located *there* in sacred space.

Drama as Immersion and Interactivity: A Very Short History

The history of drama is marked by an alternation of immersion-seeking and interaction-promoting moments. In this nonlinear evolution, the periods and movements that come closest to reconciling the two dimensions are those that remember the genre's ritual origins. Two philosophers, Aristotle and Nietzsche, have been particularly influential in imposing the view that ancient Greek tragedy derives from ritual, and though their theories are by no means regarded as unassailable, I use them as the basis of my discussion. This means that I examine *interpretations* of tragedy more than the historical institution itself.

Aristotle locates the source of Greek tragedy in the dithyramb, the sung and improvised poetry that traditionally accompanied the orgiastic cult of Dionysus (*Poetics* 3.3). This claim forms the basis of Nietzsche's classic work *The Birth of Tragedy,* an inspiring but highly personal interpretation of Greek drama whose interest resides more in its brilliant insights into the nature of art than in its historical accuracy. Nietzsche describes classical Greek tragedy as the perfect collaboration of the spirit of the two gods Dionysus and Apollo. As artistic principles, the Dionysian and the Apollonian are too complex for a straight correlation with immersion and interactivity, but the entrancing possession of the Dionysian is a prime example of immersive experience, while the Apollonian urge to capture vital forces in an artistic form requires the distanced perspective that underlies the interactive stance. Nietzsche describes the social and aesthetic function of tragedy as a taming of the horror of life through a Dionysian shattering of the individual and a joyful "fusion with primal being" (*Birth of Tragedy,* 65). By forcing the spectator to look into the abyss underlying everything, tragedy should provide the metaphysical comfort that "life is at the bottom of things, despite all the changes of appearances, indestructibly powerful and pleasurable" (59).

Aristotle takes a different view of the event that forms the ritual purpose of tragedy: "Tragedy is an imitation of an action that is admirable, complete and possesses magnitude; in language made pleasurable, each of its species separated in different parts; performed by actors, not through narration; *effecting through pity and fear the purification* [catharsis] *of such emotions*" (*Poetics* 4.1; my italics). In its attributing to tragedy a therapeutic value, the concept of *catharsis* seems self-explanatory, but its precise interpretation has raised heated controversies. Is *catharsis* synonymous with the aesthetic pleasure of tragedy, or is it a supplemental effect? Do all spectators need to be "purified" from terror and pity, or only those who are crippled by an excess of these emotions—in which case *catharsis* is more an exorcism of the sick than a moral event of universal value? How can one explain that, by eliciting terror and pity, tragedy will free the spectator from the negative effect of these emotions? Is tragedy a sacrifice and the tragic hero a scapegoat whose downfall purifies the community from harmful feelings? Or is Aristotle proposing a kind of homeopathic cure for existential anguish: experience a little dose of fear and pity in the fictional world, and you will be relieved of these feelings in the real world. Does *catharsis* reduce tragedy to a utilitarian purpose and downplay purely aesthetic pleasure,[6] or on the contrary, is the concept an acknowledgment of the scandalous fact that experiencing terror and pity for fictional individuals is a source of pleasure—the "purification" of these feelings meaning their aesthetic sublimation? And, finally, could Aristotle be telling us that the pleasure taken in tragedy is morally pure and not a matter of sadistic *jouissance* at the suffering of the hero? No matter how we resolve these individual questions, however, the concept of *catharsis* presents tragedy as the instrument of a deeply transforming spiritual event and the spectator's immersion in the fate of characters as the mode of action specific to the genre.

For all its ritual power, though, tragedy is a spectacle and not a participatory event. The play acts upon the spectators, but the spectators do not act upon the play, nor do they receive an active role in the script. Greek tragedy can be read as a reconciliation of immersion and interactivity only if we look at what happens symbolically on stage in the dialogue between the chorus and the characters. The function of the chorus is one of the most hotly debated features of ancient tragedy. One common theory, first voiced by G. F. Schlegel, interprets the chorus as the voice of an "ideal spectator." In *The Birth of Tragedy,* Nietzsche first rejects this hypothesis on the grounds that the spectator views the play as a work of art—as a fiction—while the chorus interacts with the characters as real human beings, but later he suggests that the audience identifies with the chorus (57). The chorus is not the spectator as real-world individual but his recentered counterpart in the fictional world, a counterpart deprived of agency. According to Nietzsche, the architecture of the Greek amphitheater favors this projection: "A public of spectators as we know it was unknown to the Greeks: in their theaters the terraced structure of concentric arcs made it possible for everybody to actually *overlook* the whole world of culture around him and to imagine, in absorbed contemplation, that he himself was a chorist" (*Birth of Tragedy,* 63).

Usually impersonated as a collective character (Trojan women, daughters of Oceanos, old men of Thebes), the chorus fulfills several functions—narrating the past (especially in Aeschylus), commenting on the action, weighing possible developments, or lamenting the fate of characters. For all its vocal presence, however, it has no influence on the development of the plot. The narratorial, extradiegetic function presupposes that the chorus knows more than the audience, thus precluding strict identification, but all the other types of intervention offer credible reactions from the spectator's fictional counterpart in the fictional world. In contrast to what a real audience would say, these interventions are couched in beautiful language, follow the proper metric pattern, and occur at appropriate times in the development of dramatic tension. A simulation of interactivity, the vicarious participation of the audience through the chorus is the only way to acknowledge the spectator's voice and presence without threatening the aesthetic integrity of the performance—the only way to achieve a compromise between Dionysus, who wants to lure the spectator into an individuality-threatening celebration of life, and Apollo, who wants to confine violence to the stage and disarm it through artistic form.

As responses to dramatic performance, immersion and interactivity are strongly affected by the arrangement of theatrical space. It is because standard theaters are designed for immersion, not for interaction between the actors and the audience, that the young Susanne Langer was so upset when the actress playing Peter Pan asked the children in the audience to affirm their belief in fairies (*Feeling and Form,* 318; quoted in the epigraph to this chapter). The breaking of the illusion and the loss of pleasure that ensued from this ill-fated attempt to establish an interactive relation with the audience coincided for the little girl with the moment when she became aware of hundreds of children watching the play with her.

Whereas the designs that promote immersion keep the audience hidden from itself, the quintessential interactive architecture is a circular arena, such as a sports stadium, that allows spectators to see each other as well as the action on the field (fig. 22, panel A). During a sports event, fans yell at the players, comment loudly on the quality of play, and engage in cheering duels with the partisans of the other

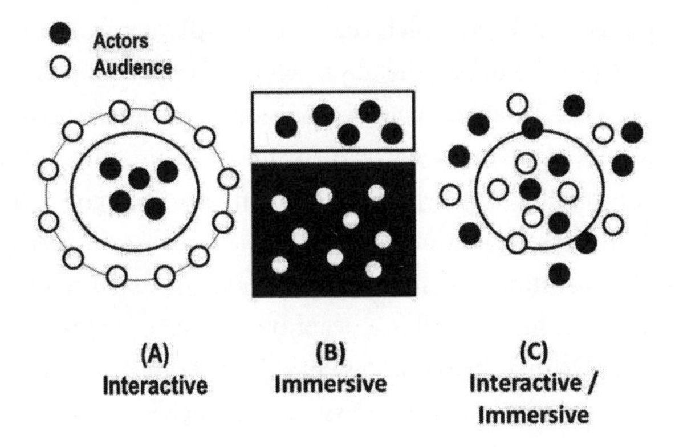

Figure 22. Three types of stage design

team. In a truly *inter*active feedback loop, the game elicits cheering, and the cheering influences the outcome of the game, creating the much hyped home field advantage. But this model of immersive interactivity is obviously not applicable to the theater. The representation of a tragedy under the circumstances of a sports event would result in a cacophony of actors and spectators talking to each other in all possible directions: actors to actors, actors to spectators, spectators to spectators, and spectators to actors. In this chaotic cross fire of addresses, the spectators would be totally distracted from the plot.

The Greek stage design, with its semicircular seating, offers a compromise between immersion and interactivity rather than a fusion. While the architecture acknowledges the audience's presence, thus establishing a spirit of communion between actors and spectators, it also maintains a strict separation between the stage and the seats, which favors make-believe. The architecture makes it clear that the audience has a role in the performance, but this role does not permit interference. A similar compromise between the interactive and immersive stances was maintained in the open-air representations of medieval mystery plays performed on the parvises of cathedrals and in Elizabethan stage architecture.

In the seventeenth century the balance of the two modes of participation was broken in favor of the immersive pole. The so-called Italian stage design of the Baroque era perfects the art of make-believe by taking the obliteration of the audience as far as it can go. Enveloped in darkness, the spectator sits in front of a brightly lit and lavishly decorated stage, imaginatively separated from the setting of the action by the transparent fourth wall and physically kept away from the raised stage by the proscenium and orchestra pit (fig. 22, panel B).[7] Walter Benjamin describes the orchestra pit as "the abyss which separates the players from the audience as it does the dead from the living; the abyss whose silence in a play heightens the sublimity" ("Epic Theater," 156). The strict division of the house between stage and auditorium does not alienate the spectator from the action but, on the contrary, heightens the illusion of live presence and authenticity. Though spectators cannot step onto the stage, they are, fictionally, part of the same world, just as the spectator of a perspective painting is included in the imaginary extension of the pictorial space. The transparent fourth wall is a peephole that enables the spectator to spy on characters who live their lives unaware of this observing presence.

Not only does the stage design suppress the audience for the actors, it also suppresses the actors for the audience. The actor of the Baroque age was not supposed to be acknowledged as a performer but to dissolve his identity into the self of the impersonated character. Writing in 1657, the French drama theorist d'Aubignac enjoined actors to act "as if there were no spectators. All the characters must speak and act as if they were really Kings, and not as Bellerose and Mondory, as if they were in the palace of Horace in Rome, and not in the hôtel de Bourgogne in Paris; and as if nobody saw nor heard them" (*Pratique du théâtre,* quoted in Rousset, *L'Intérieur,* 171; my translation). Another French theorist, Champlain, wrote in 1630 that the purpose of theatrical performance is to "surprise the imagination of the spectators" by presenting everything on the stage as "genuine and present." All difference between representation and represented must be erased. The purpose of this perfect replication is to "eliminate any opportunity for the spectators to reflect on what they are seeing and to doubt its authenticity" (quoted in ibid., 170). Illusion is reinforced through elaborate stage decoration and complex machinery—the deus ex machina is a Baroque invention—but in contrast to the medieval stage, which represented heaven and Earth simultaneously, as would a cross-section of the house of the universe, the seventeenth-century stage was entirely occupied by one spatial setting. Immersion cannot be complete if the audience's attention is divided among several locations.

Rather than inviting the recipient to a willing suspension of disbelief, as does written fiction, the theater of the Baroque age thus opted for an aggressive creation of belief through the tricking of the senses. As Rousset observes, the trademark of seventeenth-century dramaturgy is a heavy reliance on a vocabulary of magic, sorcery, waking dream, seduction, and enchantment that presupposes passivity and loss of self-control on the part of the spectator. Yet the Baroque age was acutely aware that when the curtain falls, the enchantment abruptly ends, and it maintained, as already noted, a highly ambiguous attitude toward appearances. Its dramatic magicians also knew how to practice the demystifying game of the play within the play, reminding the spectator that immersion means illusion, illusion means deception, and the beauty of their spectacles is only the fugitive beauty of dreams.

At the other end of the immersive/interactive spectrum is the epic theater of Bertolt Brecht. The aesthetics Brecht developed in support of his conception of the theater as an instrument of political action systematically inverts the values of the Baroque era and of its less playful successor, the naturalistic theater of the late nineteenth century. The famous *Verfremdungseffekt*, or V- (distancing) effect, was meant to encourage critical thinking by preventing any kind of immersion. The spectator was not supposed to regard the plot as the manifestation of an inexorable fate and to empathize with the characters—this would be equivalent to giving money to the poor, a Band-Aid solution to social injustice—but to analyze the circumstances that created their problems, to consider what other courses their lives could have taken, to assess what kind of action and attitude would be necessary to divert this course, and to apply the knowledge thus gained to real-world action.

Though Brecht's plays are often performed on a standard stage, the author's stated ambition to induce a critical evaluation of the social reality depicted in the text makes them very well suited for a circular theater. In the epic theater, temporal immersion is inhibited through songs, interruptions of play, and abstracts of the episode to come; spatial immersion through a minimalist stage setting in which signs often take the place of props; and emotional bonding through a style of acting that encourages actors to step out of character. As James Roose-Evans writes, "The actor was . . . not to 'be' Galileo but to impersonate him. To achieve this Brecht insisted that during rehearsals his actors should prefix their lines with the words 'he said' or 'she said'" (*Experimental Theatre,* 69). This distancing of the actors from their characters not only exemplified how the play should affect the audience, it was also meant to contribute to the critical education of the actors themselves: "In every instance," writes Walter Benjamin, "the epic theater is meant for the actors as much as for the spectators. The didactic play . . . facilitates and suggests the interchanges between audience and actors and vice-versa through the extreme paucity of the mechanical equipment. Every spectator is enabled to become a participant" ("Epic Theater," 154).

The term *participation* in this last sentence should not be interpreted as an invitation to the spectators to see themselves as members of the fictional world. If an actual dialogue takes place between the actors and the audience—an event that rarely happens but that would be very much in the spirit of Brechtian aesthetics—this means not that spectators step into the play but rather that actors step out of their roles to engage in a critical discussion. The locus of interaction is the real world, not the fictional one. In his theoretical thinking Brecht may, however, have underestimated the power of immersion, a power that the playwright in him knew very well how to unleash. "In one performance of *The Threepenny Opera,*" writes Roose-Evans, "it was obvious that his intentions had failed when the audience became totally absorbed in the plot, left the auditorium whistling Kurt Weill's tunes, and even found themselves identifying with Polly Peachum" (*Experimental Theatre,* 70).

Brecht's aesthetics reject immersion in favor of distanced, critical evaluation, but this critical stance can only be regarded as interactive in the metaphorical sense of the term. The audience is invited to discuss the play but not to step on the stage nor to alter the script. In the late twentieth century, the Brazilian theater director Augusto Boal adopted Brecht's anti-immersive stance and conception of the theater as an instrument of political action, but he sought to engage the audience in a much more creative way, a way that anticipates the Barthes-inspired ambition of early hypertext theorists to turn the reader into a writer.[8] "'Spectator,'" writes Boal, "is a bad word! The spectator is less than a man and it is necessary to humanize him, to restore to him his capacity of action in all its fullness" ("Theatre of the Oppressed," 352). Boal regards the poetics of Aristotle as "the poetics of oppression: the world is known, perfect or about to be perfected, and all its values are imposed on the spectators, who passively delegate power to the characters to act and think in their place. In so doing, the spectators purge themselves of their tragic flaw—that is, of something capable of changing society" (352). Catharsis, for Boal, is a purification of the spectator's desire to act; its function is the passive acceptance of the state of the world depicted in the play: "A catharsis of the revolutionary impetus is produced! Dramatic action substitutes for real action." In *Theatre of the Oppressed* (1974), Boal articulates his vision of anti-immersive interactivity through a number of street-theater projects in which the audience is invited to produce and enact scripts, rather than speaking from "the comfort of [their] seat[s]"(344). Distrusting the ambiguity of pure language, as well as the hollow rhetoric that spectators may display when publicly commenting on a play, Boal wants the audience to express itself through dramatic action because he views this action as a rehearsal of real-world political action rather than as a substitute for it, as in Aristotelian dramaturgy: "anyone may propose any solution, but it must be done on the stage, working, acting, doing things" (344). Here is an example of the kind of street event that Boal proposes to implement his vision of active audience participation. In a first stage, the actors improvise a performance with the help of a predefined script that highlights a social problem. When the action reaches a crisis, they ask the spectators to propose solutions. These solutions are enacted, either by the actors or by members of the audience. In a last stage, the solutions are collectively discussed until a consensus is reached as to which one is the best. In a nod to the virtual as potential that foreshadows the many paths of digital narrative, Boal stresses the importance of giving people "the opportunity to try out all their ideas, to rehearse all the possibilities, and to verify them in practice, that is, in theatrical practice" (345). For, "it is not the place of the theater to show the correct path, but only to offer the means by which all possible paths are examined" (ibid.). This claim may seem incompatible with a determinate political engagement such as Boal's, but it can be reconciled with his leftist agenda if one regards the multiple paths to be examined as converging toward a common goal, a goal implicitly shared by the audience and the actors that isn't—or rather shouldn't—be called into question: helping the oppressed to choose the best way to liberate themselves.

While Brecht and Boal regard immersion as a threat to political activism, a French writer and drama theorist who spent a good part of his life in mental institutions proposed a return to ritual that aims to reconcile immersion and interactivity through a trancelike involvement of the audience. Exerting an enormous influence on the experimental theater of the postwar era, the writings of Antonin Artaud (1896–1948) envision the impact of theatrical performance on the spectator as a transforming experience of almost unbearable intensity; hence, the choice of the label "theater of cruelty": "Theater is first ritualistic and magic, in other words bound to powers, based on religion, on actual beliefs, and whose effectiveness is conveyed through gestures, directly linked to the rites of theater which is the very practice and the expression of a hunger for magical and spiritual manifestations" (*Artaud on Theater,* 124). Artaud, who is credited with the first modern use of the term *virtual reality* (Hillis, "Virtual Reality," 510),[9] conceived the ritual impact of dramatic performance as a visceral experience that restores the integrity of the self through the involvement of both body and mind. The experience itself is described in rapturous tones that owe a great deal to Nietzsche's doctrine of the Dionysian essence of tragedy: "Briefly, it seems the highest possible concept of theatre is one which philosophically reconciles us with Becoming and which, through all kinds of objective situations, suggests the covert notion of the passage and metamorphosis of ideas, far more than the shock of feelings transmuted into words" (Artaud, *Artaud on Theater,* 113). Or, again, Artaud writes, "A real stage play disturbs our peace of mind, releases our repressed subconscious, drives us to a kind of potential rebellion (since it retains its full value only if it remains potential), calling for a difficult heroic attitude on the part of the assembled groups" (116).

The means Artaud recommends to achieve this painful yet healing experience anticipate many of the themes of cyberculture and many of the ideals of VR theorists. For instance:

Liberating the theater from the hegemony of spoken language and replacing the text of the author with the body-language of the actors. Modern theater must become "a new bodily language no longer based on words but on signs which emerge through the maze of gestures, postures, airborne cries . . . leaving not even the smallest area of stage space unused" (88). "I maintain that this physical language, aimed at the senses and independent of speech, must first satisfy the senses" (92).

Creating a "poetry of space" through stage design. "The stage is a tangible, physical space that needs to be filled and it ought to be allowed to speak its own concrete language" (92). Such a language should permit "spatial poetry to take the place of language poetry" (93). This view anticipates the VR developer Randall Walser's characterization of the cyberspace artist as a "space-maker" ("Spacemakers," 60).

Developing theatrical performance into a multisensory, multimedia event. "Practically speaking we want to bring back the idea of total theatre, where theatre will recapture from cinema, music-hall, the circus, and life itself, those things that always belonged to it" (*Artaud on Theater,* 109). "This difficult, complex poetry assumes many guises; first of all it assumes those expressive means usable on stage such as music, dance, plastic art, mimicry, mime, gesture, voice inflection, architecture, lighting and decor" (93).

Placing the body of the spectator at the center of the theater. "We intend to do away with stage and auditorium, replacing them by a kind of single, undivided locale without any partitions of any kind, and this will become the very scene of the action. Direct contact will be established between the audience and the show, between actors and audience, from the very fact that the audience is seated in the center of the action, is encircled and furrowed by it" (104).

Artaud's idea of a stage architecture that surrounds the spectators (fig. 22, panel C) had a great deal of influence on the avant-garde theater of the 1950s through the 1970s. In 1952, for instance, John Cage staged a happening in which the audience, sitting in the middle, could watch several activities taking place simultaneously. The 1971 production of the historical spectacle *1789* by the Théâtre du Soleil, directed by Ariane Mnouchkine, placed the audience on a platform surrounded by scaffolding on which actors performed song, dance, and mime, using masks and puppets and without the guidance of an authorial text (Roose-Evans, *Experimental*

Theatre, 75 and 88–89; see also Popper, *Art,* 120–21). The encircling design offered a visual metaphor for immersion, and the disappearance of the proscenium made it physically possible for the audience to mingle with the actors, but most of these projects did not permit a literally productive participation of the spectators in the play. It was still the actors who performed and the spectators who watched.

The caveat of letting spectators on stage is that if they gain control of the action, the resulting performance may become utter chaos, since, unlike actors, spectators are not trained to perform a script. If only a small fraction of the public is allowed to participate, the majority will remain in a noninteracting role, and they will have to put up with the unpredictable and, most likely, unpolished contributions of the selected few. On the other hand, if everybody is let on stage, how can interaction among so many people be coordinated into a performance that will be pleasurable for all? As Brenda Laurel observes, the idea of the audience as active participant is more likely to add "clutter, both psychological and physical" to the stage than to lead to an artistic event (*Computers as Theatre,* 17). To circumvent this problem, Artaud's theater of cruelty resorts to a vicarious interactivity, as did Greek drama before it. It is through a communion, almost a transubstantiating identification, of the spectators with the actors that the performance exercises its ritual purpose of a "reconciliation with Becoming." In some participatory productions—for instance, *The Domestic Resurrection Circus* by the Bread and Puppet Theatre, 1970 (Roose-Evans, *Experimental Theatre,* 123)—the audience is invited to invade the stage at the end of the show and to join the actors in a chant or song, but this literal communion concludes a ritual in which the actors remain very much in control of the action, as do the priests in a religious ceremony before the Holy Spirit descends on the congregation.

The alternative to vicarious interactivity is to coach the audience into taking part in a disciplined action, so that literally there will be no spectators, and the play will be staged for the benefit of its own participants. In this situation, as Laurel puts it, "*the representation is all there is*" (*Computers as Theatre,* 17; italics original). We have seen this idea of the actors becoming their own audience, which inverts the concept of the spectator turned actor, surface in the discussion of Brecht's dramatic philosophy. It flourished in the political street theater of the Vietnam War and civil rights era, when performance groups such as the Living Theatre of Julian Beck and Judith Malina approached the theater as a way of life, adopted a communal lifestyle, developed their spectacles collectively, and regarded them as a means of self-discovery. "The [members of the] Living Theatre," wrote Peter Brooks, "are in search of a meaning in their lives, and in a sense even if there were no audiences they still would have to perform because the theatrical event is the climax and the centre of their lives" (quoted in Roose-Evans, *Experimental Theatre,* 104).

When performing becomes synonymous with living, the theatrical experience inherits the immersive and interactive qualities that define our experience of being-in-the-world. This fusion of life and representation and this total engagement of the actors are, of course, far too utopian to offer any kind of useful guidelines for the developers of electronic forms of interactive art. Nor is it practical to coach a large number of participants to perform a script, even one that allows ample room for creative improvisation. These difficulties leave only one viable option for the design of immersive/interactive experiences: let only one or two users, at the very most a handful, into the virtual world. This ability to customize virtual worlds and artistic experiences to individual users is precisely the forte of electronic media.

Tabletop Role-Playing Games

In all the examples I have discussed so far, interactivity is either understood in a metaphorical way or severely limited by a predefined script. It is in the genre of the tabletop role-playing game, which originated in 1974 with Gary Gygax and David Arneson's famous game Dungeons and Dragons, that active user participation and narrative immersion achieve the highest degree of compatibility. As their name suggests, tabletop role playing games (henceforth RPGs) are played by a group of players who come together in real space and play the game with physical tokens, such as dice and a rule book. The games involve the following fictional and real elements: an imaginary game-world, created by a story; a number of participants who play the roles of characters; a game master, who directs the unfolding of the game from a position external to the game-world; a narrative script described in a game book, capable of many variations that the players enact under the guidance of the game master; and a set of rules, also described in the game book, corresponding to the actions through which the characters can interact with the world.

As Susana Tosca and Olivier Caïra argue, the plots of RPGs are neither linear nor arborescent; rather, they emerge dynamically during gameplay, out of the possibilities associated with the various components of the game-world. Places, characters, and objects all come with specific affordances, which determine what can be done in a given situation. Here is an invented example of a strategic situation:

> You are lost in the forest; it is getting darker and darker and it is starting to snow; you have no place to sleep nor any warm clothes; but suddenly you see a faint light on the horizon: a house! Maybe the people can put you up for the night.

> As you get close, however, you see that there is a large river between you and the house. What do you do?

Players may respond with the following action:

> You cast a spell to dry up the river so you can cross it [dice throw], but now you are out of magical aides and you cannot restore the flow. The consequences for the inhabitants of the region will be disastrous. When they find out who dried up the river, you'll be in serious trouble.

The game unfolds as an alternation of situation description by the game master (based on the narrative script of the game book), discussion of possible action by the players, action taking through the throw of the dice by individual players, and narration of the result and consequences by the game master. Here is an example of live interaction (recorded and quoted by Caïra, *Jeux de rôles,* 115; my translation).

> *[The players, having discovered that a NPC named Laryon is a traitor, decide to attack him]*
>
> FLORIS: *[after passing up a turn in the fight]*: OK, I'll try the sabre. *[Rolls a 96]*
>
> GAME MASTER: Critical failure: your sabre gets stuck in the staircase railing, and it breaks when you try to free it.
>
> JULIEN: I'll take my turn.
>
> GM: *[to Floris]* Now you have no weapon left. He *[Laryon]* hits you. *[Dice roll].* The blow partly misses you, but you still have to cower.
>
> FLORIS: *[Dice roll.]* Rats, I needed 55, and I got 56. *[Laughs]*
>
> GM: You lost two life points.
>
> EMMANUEL: I'll try a spike in the eye.
>
> GM: You must first try: "stab the guy"; then we'll see if you have enough control to hit the eye.

In the above example, the game master speaks either as third-person narrator ("your sabre gets stuck") or out-of-role ("you must first try"). In the next example (describing the same fight, a few blows later), the game master also impersonates the NPC Laryon. The players join in the dramatic enactment, directly answering Laryon.

> GM: And his counterblow fails miserably. *[Playing the hurt Laryon]* "Aarg. Curse on you. You are not simple merchants. Who are you?"
>
> JULIEN: "Simple merchants, with a life behind them."
>
> EMMANUEL: "Simple merchants, with whom you tried to deal."

GM: "People who will not speak. You are right: your life is behind you." *[Floris laughs.]* He tries to hit you with a broad strike. *[Dice throw]* The blade just misses your head.

Not all possible developments need to be described in the rule book. Thanks to her natural intelligence, the game master can make new rules on the fly when the players take the story in an unforeseen direction, or she can overrule the book, when she can imagine more exciting developments. The participants' enjoyment depends not only on the completion of a coherent story arc but also on the leadership and storytelling talent of the game master, as well as on the players' ability to give life to their characters through their improvised enactment. Pleasure is as much imaginative as it is strategic, as much a matter of immersion in a fictional world as a matter of successful problem solving.

RPGs introduced many features that have since become a standard part of computer games. For instance, in the original Dungeons and Dragons, players could determine the properties of their characters, choosing among seven races with distinct characteristics (human, dwarf, elf, halfling, gnome, half-orc, half-elf), and four social classes with different abilities (fighter, magic user, cleric, and thief). The game was organized into different levels, and characters had to restore their health through rest or magic potions when their "health meter" reached a critical low as the result of fights or strenuous efforts (Caïra and Tosca, "Role-Playing Games," 433). But while computer games have learned important mechanisms from RPGs, they have not been able to emulate the emergent quality of RPG stories because no synthetic agents can match the ability of naturally intelligent players to adapt a story to dynamically evolving, unpredictable situations. We will see in the next chapter what features computer games can contribute in the pursuit of an immersive and interactive narrativity.

TEN

Chasing the Dream of the Immersive, Interactive Narrative

Cyberculture and postmodern theory have popularized the view that we own not simply a physical body—given to us, mortal, subject to irreversible changes, limited in its abilities, and anchored in "real reality"—but also numerous virtual bodies, or body images, which either clothe, expand, interpret, hide, or replace the physical body, and which we constantly create, project, animate, and present to others. If the artistic reconciliation of immersion and interactivity requires the participation of the body in the art world, there is no reason why the bodies in question could not be of the virtual kind. Virtual bodies are implicated in numerous human experiences, and they entertain various relationships to the physical body. Some, like the bodies impersonated by dancers and actors, follow exactly the movements of the physical body but express a totally different self, while others, like the bodies that we enter in dreams, belong to the same self as does the physical body (it is still *me,* though my dream body can fly), but they are not animated by the same voluntary muscles. Still other body images are operated by the physical body, but the gestures of the physical body do not correspond to those of the virtual one: while one body slays dragons, flirts with an insurance salesman who poses as a hooker, or explores an enchanted forest, the other one types on a keyboard or squeezes a joystick. Such are the bodies that we take across the screen to reach imaginary worlds in standard two-dimensional electronic media.

Since our embodied relation to the life-world is both immersive and interactive, the most intense combination of immersion and interactivity occurs in those online worlds that simulate life experience. A genealogy of immersive/interactive worlds leads from Dungeons and Dragons, the popular tabletop role-playing game of the 1970s, to the text-based MUDs (multiuser dungeons) and MOOs (multiuser domains, object-oriented) of the 1990s, in which users do not play a structured game but freely interact under self-constructed make-believe identities; and from MOOs and MUDs to the visually rendered online worlds of today, such as MMORPGs (massively multiplayer online role-playing games), in which players not only engage in "quests" with a fixed narrative script but also socialize, explore, and acquire commodities, and Second Life, an initially empty world that users fill with their own creations, including their avatars. In contrast to MMORPG players, Second Life users do not perform quests with a fixed script but rather enact the life story of their avatars by building, by buying and selling things, by manipulating the objects that fill the world, and above all, by forming relationships with the avatars of other players. The literature on Second Life (and before, on MOOs) is full of expressions of fanatical loyalty on the part of its users (Boellstorff, *Coming of Age*). Writing in the 1990s, Sherry Turkle reported several cases of MOO users who regarded their MOO identities as "more real" than their ROL (rest-of-life) selves, presumably because they took greater pleasure in these masquerades than in face-to-face interaction with flesh and blood individuals. These remarks have lost none of their validity in today's multimodal worlds. But for all their potential immersivity and interactivity, online worlds cannot be regarded as a productive formula for immersive/interactive narrativity because, in the words of Elizabeth Reid (written about MOOs), "The system provides players with a stage, but it does not provide them with a script" ("Virtual Worlds"). Whereas in MMORPGs the system tells people what to do (though they are free to do something else), in MOOS and Second Life, players are entirely responsible for their own entertainment. The narrative (or should one say dramatic) quality of the players' experience depends on who happens to be online, on whether they have formed rich imaginative relationships with other players, and, more literally than in any other mode of communication, on their own performance. A visit to Second Life, or to a MOO, may or may not result in a virtual life experience that the user will remember as a story.

In this chapter, by contrast, I look at attempts to create truly *designed* narrative experiences that combine immersion with interactivity. The question I ask is not "how to create an environment that is both interactive and immersive," since both single-user and multiuser video games can fulfill these requirements, but rather how to make narrative interactive without losing its proven immersive power—its narrativity. As I noted in chapter 3, immersivity can be understood in two ways: in a properly VR sense, as the technology-induced experience of being surrounded by data, and in a narrative sense (which is compatible with the other kind), as being imaginatively captivated by a storyworld. I begin my investigation with an example of technological immersion, *Placeholder*, by Brenda Laurel, Rachel Strickland, and Rob Tow, an early VR installation project that shows how the whole of the physical body can be involved in a script that possesses some degree of narrativity.[1] Then I focus on screen-based computer games and playable stories as instances of interactive narratives that aim at the imaginative kind of immersion.

Digital Installation Art as Interactive

The 1990s saw not only the popularization of the idea of virtual reality but also the development of art installations that relied on full-body immersion in a fully or partly simulated world. These installations have various themes and purposes, but most of them explore the role of the body in the experience of the world. *Placeholder* stands out from other projects through its compelling dramatization of embodied perception.

Placeholder grew out of Laurel's fascination with the affinities between VR and ritual. The installation is an electronic demonstration of her belief that VR can fulfill the same spiritual function of fostering communion with life forces as did Dionysian festivals and, in her speculations, the ceremonies of the great kivas of the Anasazi civilization.[2] "Virtual reality may be many things," she writes. "It may become a tool, a game machine, or just a mutant form of TV. But for virtual reality to fulfill its highest potential, we must reinvent the sacred spaces where we collaborate with reality in order to transform it and ourselves" (*Computers*, 197). The script of *Placeholder* enacts the ritual pattern of penetrating into a magic circle, being reborn in a different body, and acquiring enhanced powers of perception that deepen the bond of the subject to the natural world.

The project uses three-dimensional videographic scene elements, spatialized sounds and words, and simple character animation. Two users, equipped with head-mounted displays, wander for about fifteen minutes—though some of them took almost two hours—inside a ten-foot diameter area, the "magic circle," which corresponds to the range of the equipment that tracks their movements and generates the visual and auditory display. The two participants are said to be "physically remote" (Laurel et al., "Placeholder," 118), and their activity consists primarily of an individual exploration of the virtual world, but insofar as they can hear and see each other's character, they engage in a rudimentary form of dramatic interaction.[3] The system does not need AI components—all texts are prerecorded—but one character is improvised live by an actor. This character, the "Goddess," functions as online help, initiating users into the secrets of the virtual world and providing internally an equivalent of the knowledge that children bring into games of make-believe.

In stark contrast to the Aristotelian design philosophy advocated in Laurel's book *Computers as Theatre,* the narrative structure of *Placeholder* is not dramatic but predominantly epic. The system's openly stated purpose is to create a sense of place, and its architecture is more indebted to a poetics of space than to a poetics of plot. The sense of place is conceived as an encounter of the user with what the Romans called the genius loci, a spirit that protects a site and safeguards its essence. The spirit of a place manifests itself through the narratives that recount its origin, thereby establishing its sanctity and affective significance: "When a person visits a place, the stories that are told about it—by companions, by rock art or graffiti, or even by oneself through memories or fantasies—become part of the character of the place" (121). To dramatize this idea, *Placeholder* creates a magical space inhabited by animated objects. The user wanders around an environment teeming with narratives in which every creature and every landscape feature has a story to tell.

The setting of *Placeholder* consists of three distinct sites inspired by actual locations found near Banff National Park: a hot spring in a natural cave, a waterfall, and a landscape of rock formations (hoodoos) created by erosion. Dark portals separate these different locations, so that moving from one to another is similar to undergoing a rite of passage. The world of *Placeholder* is inhabited by four mythical "critters": Spider, Snake, Fish, and Crow. As the user enters the virtual world, these critters are represented as petroglyphs on the walls. The user cannot see anything, but she hears faint voices that emanate from the petroglyphs; these are the voices of the critters speaking about themselves. Fascinated by their narratives, the user comes closer and hears the story more and more distinctly. In a forceful allegory of the immersive power of narrative, the user who crosses a certain threshold toward a critter becomes embodied as that critter, suddenly taking on its appearance, voice, point of view, and mode of locomotion. Now she is able to see the world from a different perspective: Crow's vision captures spectacular reflections, Spider has eight eyes corresponding to eight point of views, Snake can see in the dark, and Fish sees underwater.[4] The system enhances the user's sense of identity with her reconfigured body by making it visible to her and to others, but here acquiring a virtual body is not merely a matter of presenting it to others, as it is in online worlds such as Second Life. It is mainly by inhabiting her new body from the inside and by reaching with it toward the world that the user acquires "a new sense of what it is to be an embodied [creature]" (125). Through the theme of metamorphosis, the developers of *Placeholder* hope to increase the user's sense of embodiment, not only by altering her mode of perception but also by making her construct her own body through the action of moving toward a critter. The theme of corporeal and perceptual rebirth is underscored by the fact that as long as users are human, they can only see their hands; but once they become reincarnated as one of the critters, their full body image becomes visible to them and to the other participants. Another function of the metamorphosis is to enhance the user's verbal creativity. By giving her a new body, the system also gives her a persona to act out and experiences to talk about. Users leave their mark on the virtual world by interacting verbally with it. Their utterances are recorded and stored in devices called "voice holders." These devices, actually digital images that look like rocks, can be transported around by a handheld device (much in the way a mouse drags icons on the computer screen) and replayed by the next users. Interactivity is not merely a matter of walking, speaking, and using hands to touch and move objects around, but more importantly it is a matter of enriching the narrative tradition that expresses and creates the spirit of the place. It reaches symbolically across time and generations, as users talk to their successors and listen to their predecessors. At least, they do so in the vision of *Placeholder*'s creators. It would take an experimental study of audience behavior to tell whether the visitors had interesting stories to tell and whether they had the patience to listen to the stories of their predecessors before they started exploring the world through their new body.

Because of technological (and financial) limitations, *Placeholder* did not live up to the ambitions of its creators. Some of the "critters" were never implemented, and lag between the user's actions and the response of the system was a deterrent to immersion. So was the heavy machinery necessary to the installation. Far from being specific to *Placeholder*, this technological overload clouds the future of digital installation art as a vehicle of interactive narrative. The equipment is expensive, highly specialized, and usually clumsy, and since installations are normally part of a public display, severe limitations must be placed on the duration of the visit for the work to allow a reasonably efficient throughput. These time limits are not conducive to narrative development, since stories require a spacious time frame. As *Placeholder* demonstrates, it is fairly easy for digital installations to generate spatial immersion, but in order to provide a truly narrative experience, these projects will have to do more than let the user loose in a simulated world—they will have to generate temporal and emotional immersion. We will have to await the creation and commercialization of affordable personal VR systems to achieve full-body immersion in storyworlds, but the hardware problems pale in comparison to the AI and design problems that remain to be solved in order to overcome the interactive paradox discussed in chapter 7. I take a closer look at these problems and at some of their solutions in the next sections.

Narrative Games and Playable Stories

While narrativity is a type of meaning, interactivity—when it is put in the service of entertainment—is a type of play. The combination of narrativity and interactivity in 2D screen-based media oscillates between two forms: the *narrative game,* in which narrative meaning is subordinated to the player's actions, and the *playable story,*[5] in which the player's actions are subordinated to narrative meaning.[6] Or, to put it differently, in a narrative game, story is meant to enhance gameplay, while in a playable story, gameplay is meant to produce a story. The concepts of narrative game and playable story reflect, in their opposition, the distinction made by the French sociologist Roger Caillois between two types of game: *ludus* and *paidia* (*Man, Play and Games,* 13). The best example of *paidia* games is building imaginary scenarios with toys, using them, in the words of Kendall Walton (*Mimesis as Make-Believe,* 21–24), as "props in a game of make-believe." These games do not aim at a specific goal, and they do not lead to losing or winning. The pleasures of *paidia* reside in the free play of the imagination, in adopting foreign identities, in forming social relations, in building objects, in exploring an environment, and above all in creating a representation: *paidia* games are fundamentally mimetic activities. If there are rules, they are spontaneously created by the participants (as when a group of children decides that a certain tree will be the house of the wolf), and they can be renegotiated on the fly. *Ludus* games, by contrast, are strictly controlled by preexisting rules accepted by the participants as part of a basic game contract, they lead to clearly defined states of winning or losing, and their pleasure resides in the thrill of competition and in the satisfaction of solving problems.

What I call a narrative game is a *ludus* activity. If there is one contribution that digital technology has made to the design of games, it is their narrativization. By this term I mean the transformation of what used to be abstract playing fields (such as a chessboard or a baseball diamond) into concrete fictional worlds populated by recognizable objects and individuated characters. The main difference between an abstract game like soccer or chess and a narrativized video game like Half-Life, Max Payne, or Grand Theft Auto is that in an abstract game the players' goals are only made desirable by the rules of the game (who would, in real life, be interested in kicking a ball into a net or in moving tokens on a board?), while in a narrativized game the player pursues the kind of goals that people may form in everyday life or in their fantasies—goals such as saving the world from invaders and rescuing people in danger, if you want to be a good guy, or stealing cars and killing people, if you prefer villainous roles. But in the intensity of the action, players may forget whether they are terrorists or counterterrorists, space aliens or defenders of the earth; in a narrative game, the player plays to win, to beat the game, and story is mostly a lure into the game-world.

While *ludus* inspires narrative games, the spirit of *paidia* permeates playable stories. In a playable story there is no winning or losing; the player's purpose is not to beat the game but to observe the evolution of the storyworld. Playable stories induce a much more aesthetic pleasure than narrative games because the player is not narrowly focused on personal goals. For me the essence of the playable story is captured by what I once heard a little girl say about The Sims: "Guess what I managed to do with my Sims. I made the father and mother drown in the pool, and now the kids are alone in the house and they can do whatever they want." For this little girl, the pleasure of the game did not come from reaching a state defined by rules but in coaxing a good story out of the system.

The Poetics of Interactive Narrativity

Whether or not interactive narratives practically exist or are still chimeras depends on what is expected of the user's participation; the more active and the less constrained the user's role—in other words, the more lifelike, though life is not free of constraints—the more problematic its integration into a well-formed narrative arc. For intensity of user participation, freedom of choice, and depth of immersion, nothing can beat the imaginary Holodeck of the TV series *Star Trek,* which Janet Murray, in her 1997 classic *Hamlet on the Holodeck,* proposed as the model of the new kind of narrative experience that digital technology would make possible. The Holodeck is a computer-generated, three-dimensional simulation of a fictional world. The user is invited to step into this world, to impersonate a character, and to interact through language and gestures with synthetic (i.e., computer-created) agents. No matter what the user says or does, the synthetic agents respond coherently and integrate the user's input into a narrative arc that sustains interest.

As a model of interactive narrative, the Holodeck may be, in the words of Brenda Laurel, "an elusive unicorn we can imagine but have yet to capture" (*Utopian Enterpreneur,* 72), but this does not take anything away from the validity of its individual features as goals to pursue for developers of interactive narrative. In this section I propose to use its features as a point of departure for a poetics of interactive narrativity, measuring these features against the devices, interfaces, and design philosophies of actually implemented forms of narrative games and playable stories.

Natural Interface

Visitors to the Holodeck interact with the computer-generated world in exactly the same way people interact with the real world—through language and through the gestures of the body. This double means of interaction is essential to real world social life and to its narrative representation (or its computer simulation). Our relations to our physical environment consist mainly of corporeal actions, while our direct relations to other human beings are based on semiotic transactions, though we can also affect other people indirectly by performing physical actions that will please or displease them. While narrative involves both kinds of action, the interpersonal/language-based kind is the glue that ties together the physical actions. A car chase by itself may be visually stunning, but it only becomes narratively meaningful if the chaser and the chasee have reasons to behave the way they do, and these reasons can only be established through some prior acts of verbal communication, such as promising, threatening, concluding an alliance, or informing somebody of somebody else's plans. Narrative genres may differ from each other in the prominence they give to physical action and in the complexity of the interpersonal relations that motivate these actions, but extremely rare are the stories that focus entirely on problem-solving through physical actions. (*Robinson Crusoe* may be an exception, until the hero meets Man Friday.) To rival the rich narrative affordances of media such as literature and film, a system of interactive storytelling should be able to stage physical actions that change the fictional world as well as verbal acts that affect the minds of its inhabitants and motivates them to take action.

Both natural language comprehension and whole-body interface have been used in new media art and entertainment but with limited narrative efficiency. Interaction through natural language gives the user a life-like freedom of expression, but there is no guarantee that the system will understand the input and respond in a logically (and narratively) coherent ways, as the very limited success of computers in passing the Turing test demonstrates.[7] Natural language parsing has been used in the interactive drama Façade, but it is only because the user plays the role of a relatively disempowered witness in a domestic fight between the synthetic characters that she can accept the system's frequent inability to understand what she is trying to say: inconsistencies can be attributed to the uncooperativeness inherent to a quarrel and to the self-centered nature of the synthetic characters. Another form of digital narrative that uses natural language interface is interactive fiction (IF), a purely text-based genre defined by Nick Montfort as "a program that simulates a world, understands natural language from an interactor, and provides a textual reply based on events in this world" ("Interactive Fiction," 316). In contrast to Façade, which allows the user to type whatever she wants but most of the time fails to respond adequately, the IF parser sacrifices the naturalness of language by putting severe limitations on the player's use of syntax and vocabulary. Most of the acceptable inputs are two-word sentences (like "take knife"), and the parser openly rejects any input that cannot be processed. This means that the player must learn the idiom of the system as a code within a code—almost like a second language. The most extreme limitation in the use of language occurs in menu-based communication. Many computer games (for instance, Morrowind) allow the player to dialogue with the characters by selecting an item from a list of canned utterances. Here the user does not have to learn a specialized idiom, and the system is able to respond coherently to any choice, but the price to pay is a loss of fluidity, since narrative time must stop until a choice is made. The menu also clutters up the screen with an extra window that takes a toll on the player's immersion in the fictional world.

Whole-body participation, the other distinctive feature of Holodeck interaction, is frequent in digital installation art and in sports games, such as bowling or tennis simulations or in dancing games. These systems track the motions of the user's body and use this data in a variety of ways; art installations may project an image of the body in a computer-generated display so that the users can see themselves in the virtual world, while sports games calculate and display the result of the player's actions. But sports games and most digital art installations are limited to a physical interaction with the world, and they do not represent the interpersonal relations that impart narrativity to a designed experience. The alternative to full-body interaction is the manipulation of a control pad or of the keyboard. In contrast to full-body interface, this mode of interaction rests on a non-iconic, arbitrary relation between the gestures of the body in the real world and the events triggered by these gestures in the fictional world. The range of actions that can be performed by manipulating controls without interrupting the unfolding of the events is again limited to physical action, such as moving the player's virtual body, picking up objects, and firing weapons. None of these actions involves interpersonal relations, not even when players kill enemies in video games. Even though these enemies may take the appearance of human beings, from the player's point of view they are mere bodies that need to be eliminated. Only menus based on language offer the player the possibility of performing actions that affect other characters as persons, this is to say, emotionally. In The Sims, for instance, menus allow players to choose among actions such as "appreciate," "irritate," "dance with," "play," "hug," "kiss," and "talk to," all verbs that take a human being as object. The choice of action affects the emotional relations between the two characters by making the barometer of their mutual affection rise or drop.

With both verbal and corporeal action, then, interactive story systems have to choose between reasonably natural but narratively limited modes of action and artificial interfaces that suspend game and narrative time but allow much richer interpersonal relations.

Integration of User Action within the Story

Just as in real life all of our actions contribute to our life story, in the Holodeck, all of the visitor's actions move the plot forward. This degree of integration is almost automatic in a menu-driven system like The Sims, since the player's selection counts as the performance in the fictional world of the action described by words on the menu. The creation of a narrative out of the user's choices is facilitated by the nature of the options; all the items on the menu involve interpersonal relations, and the succession of choices writes the life story of the Sims family. In a system with a natural-language understanding system, the degree of integration depends on the parser's efficiency. It was my feeling while playing Façade that much of what I typed was not part of the dialogue and, consequently, of the drama, because the characters simply ignored my input. But how can a story be created when the user's possibilities of action are limited to moving, picking up objects, manipulating them, and solving riddles through this manipulation, as is the case in shooters and adventure games? The most obvious way to handle this problem is to choose a type of plot that puts great emphasis on physical actions. More intricate plots and backstories can be created through the use of film clips (known in game jargon as cut scenes) but usually at the cost of the integration of the user's actions within the narrative. In the worst of cases, the story only moves forward in those moments when control is taken away from the player, and the player's actions are nothing more than the means to unlock the next episode by solving problems gratuitously thrown along the way to give him something to do. In Myst, for instance, the player needs to pull levers, turn dials, find keys, and guess secret codes to be admitted to the next space, where she will find another page of the book that tells the past story of the fictional world. The game designer Chris Crawford calls this situation a "constipated story" (*Chris Crawford on Interactive Storytelling,* 130), and the game critic Steven Poole wittily describes it as follows: "It is as if you were reading a novel and being forced by some jocund imp at the end of each chapter to go and win a game of table tennis before being allowed to get back to the story" (*Trigger Happy,* 109).

Frequent Interaction

In real life, we are always doing something, except when we are sleeping—and maybe even then, since we remain active in dreams. For an interactive system to offer a reasonably believable simulation of life, it must therefore maintain the user in a constant state of possible action. As an example of interactive narrative that violates this guideline, consider *I'm Your Man,* a 1992 interactive movie. The user, whose involvement oscillates between external-ontological and external-exploratory, can determine whether Jack will be a hero or a coward, whether Leslie will be a good girl or a bad girl (both ontological decisions), or whether to follow Jack or Leslie (an exploratory decision), but the forking points that require action are separated by long stretches of passive viewing. In most computer games, by contrast, activity-requiring scenes are the norm, and passive ones (cut scenes) the exception. Even when players are not engaged in narratively significant actions, such as fighting enemies, they can remain active by exploring the game-world or, in a multiplayer game, by socializing with other players. In a well-designed game, there is no such thing as waiting in place until something happens.

Dynamic Creation of the Story

In the Holodeck narrative machine, every action of the visitor affects the life of his fictional persona, and every different choice leads to a different story. It would be impossible to store in advance all the consequences of all the decisions that can be made by the player. The only way for the system to deal efficiently with the visitor's freedom of action is therefore to compute the effects of her actions in real time, amending its model of the fictional world and responding to the player on the basis of this updated model. This dynamic process is known as a simulation, and it imparts to its output a quality of emergence.

Existing forms of interactive narrative can be broadly divided into bottom-up, emergent systems that create stories during the run of the program, as does the Holodeck, and top-down systems that rely on pre-scripted content. The former can be played many times, with different results, while the latter are meant for a single traversal, since the story does not renew itself.

The bottom-up approach is illustrated by The Sims. The program creates a world full of characters and things. Each of these objects is linked to a set of possible behaviors, listed on a menu that comes to the screen when the user decides to play with this particular object. When a behavior is selected, it leads to another state of the fictional world, and another set of behaviors becomes available. As the world passes from one state to another, a story is created. In this kind of system, the choices offered to the player are too numerous and the interactions of the various objects too complex for the designer to anticipate all the possible developments. Bottom-up narrative design is a little bit like the TV show *Survivor*: you throw a number of characters with well-defined personality features together in a space, and you wait to see what kind of story will come out of their interactions. If there is a drawback to bottom-up systems, it is the lack of closure of their output. Without top-down authorial control, it is virtually impossible to create a dramatic curve of rise and fall in tension or a sequence of events that stops after a conflict has been resolved. But closure is not indispensable to narrative pleasure. Throughout literary history, from the never-ending Renaissance narrative of *Orlando Furioso* to the feuilleton novels of Dickens, Trollope, or Eugene Sue in the nineteenth century, and to the modern TV soap operas, readers have time and again been fascinated by narratives that go on and on, like life itself.

While the bottom-up approach is favored by playable stories, the top-down approach is typical of narrative games, such as shooters and adventure games. In this approach, there is no event generation on the fly. The player's progression is a journey along a path that is already traced and that leads to a fixed destination or to several destinations when the system offers branching points. There are two ways to create top-down interactive narrativity. The most common is to start from a set of problems to solve, actions to take, weapons to use, effects to create—in short, starting from the design of gameplay—and to wrap this gameplay into a story. This is how for instance the game Prince of Persia was created (Mechner, "The Sands of Time," 2007). The other method consists of starting from a specific storyworld and inserting possibilities of user action to make it interactive. We see this approach in games based on *Harry Potter, The Matrix, Lord of the Rings,* or *Alice in Wonderland.* But because the plot of these games must be adapted to the possibilities of action offered by game controls, it is usually fairly different from its literary or cinematic source. Many of the games based on a preexisting story tend to become stereotyped shooters and quests, with weak integration of the player's actions into the storyline. These games attract players much more for the spatial and visual pleasure of finding themselves in a familiar fictional world and of encountering favorite characters than for the temporal pleasure of enacting a specific sequence of events. In this kind of design, storyworld takes precedence over story.

The top-down and the bottom-up approaches are not mutually exclusive.

Scripted elements can be used in bottom-up systems to give proper narrative form to the output, while top-down systems, as already noted, would not be interactive if they did not find a way to integrate the bottom-up input of the user in their narrative arc. The Sims, for instance, sparks interest by occasionally taking control away from the player in order to stage pre-scripted scenarios that create unexpected turns of events (such as a male character being kidnapped by space aliens and returning pregnant), while Façade, a basically top-down design in which the system-created characters take command of the plot and bring it toward closure, manages nevertheless to make the dialogue vary with every performance, thanks to the player's active participation. Any future solution to the paradox of interactive narrativity will lie in a novel combination of top-down and bottom-up design.

Plot Types for Interactive Narrative

The design of an interactive narrative begins with the choice of a type of story. In this section, I examine the potential of four kinds of literary plot for interactive implementation: the epic plot, the epistemic plot, the dramatic plot, and the soap opera.

Aristotle recognized two forms of narrative, the epic and the dramatic. Although the distinction was primarily based on the mode of presentation—the epic representing events through verbal narration (*diegesis*) and the dramatic through an imitation of action (*mimesis*)—the two genres also differ from each other through the form and content of the plot. As Aristotle wrote, "one should not compose a tragedy out of a body of materials which would serve for an epic—by which I mean one that contains a multiplicity of stories" (*Poetics* 8.7, 30).

The epic plot is focused on the exploits of a solitary hero; it "preserves the memory of glorious deeds performed by superior beings" who "show their mettle in battles against human foes, monsters, or the powers of nature" (De Jong, "Epic," 139). Since every feat adds to the glory of the hero, the story can be endlessly expanded by adding new feats and new episodes. Epic narratives focus on physical actions, and the human relations that motivate the hero to act remain fairly simple. Take the archetypal plot of the fairy tale, as described by Vladimir Propp in *Morphology of the Folk-tale*: a villain causes harm to a family, typically by kidnapping a princess; the hero is dispatched to repair the situation; after a certain number of tests he fulfills his mission by defeating the villain with the help of a donor, and he is rewarded for his actions with the hand of the princess. Throughout this plot, there is no evolution in personal relations. The hero is the faithful servant of the dispatcher and remains opposed to the villain until the end (there is no reconciliation). Nobody changes sides during the fight between the two factions, and when the hero marries the princess, they live happily ever after.

Dramatic narratives, by contrast, focus on evolving networks of human relations. Here are some examples of this evolution: (1) in the beginning, *x* is allied with *y*; then *x* betrays *y* and sides with *z*, and in the end *x* and *y* are mortal enemies; (2) in the beginning, *x* has always been faithful to *y*; then *x* falls in love with *z*; *y* becomes jealous and kills *x* and *z*; (3) in the beginning *x* and *y* are friends, then *y* insults *x* and they are enemies, but in the end *x* redeems himself, *y* pardons *x*, and they are friends again. In a dramatic plot, the action is mental rather than physical; most of the events consist of acts of verbal communication between the characters. When the characters perform physical actions, the significance of these actions resides in what they reveal about the mind of the agent and in how they affect interpersonal relations. Another difference from epic plots is that dramatic narratives present a closed pattern of exposition, complication, and crisis and resolution (also known as the Freytag triangle) that defies expansion. The dramatic plot's focus on interpersonal relations describes both the tragic and the comic genres.[8]

A third kind of narrative made its appearance in the nineteenth century—the epistemic narrative, driven by the desire to know. Its standard representative is the mystery story. The trademark of the epistemic plot is the superposition of two stories, one constituted by the events that took place in the past, and the other by the investigation that leads to their discovery. While the seed of the genre can be found in earlier forms of narrative—for instance, in Sophocles's *Oedipus Rex*—Walter Ong attributes its emergence to the invention of print. The intellectual appeal of the mystery story lies in challenging the reader to find the solution before it is given out by the narrative; in order to do so, the reader needs to sort out the clues from the accidental facts and to submit these clues to the logical operations of deduction and induction. This mental activity would not be possible if the print medium did not give the reader the opportunity to parse the story at her own pace.

Thanks to the unlimited time resources of television, yet another type of plot developed in the twentieth century: the soap opera, a narrative that can extend over decades. While the soap has a set beginning and end, its end is not dictated by internal necessity, which means by the completion of a narrative arc, but rather by external circumstances, such as a shrinking audience or a favorite actor leaving the show. As a serial form, soap operas are written on the fly, only a few episodes ahead of the current installment. They follow the entangled destinies of a large number of characters, alternating between multiple parallel plot lines, as shown in figure 16.

New subplots are continually created, as characters enter into new relationships or new characters are added to the cast.

Interactivity and the Epic Plot

The epic plot has long been a favorite of game designers. We find it in shooters, in adventure games, and in the quests of MMORPGs. It is easy to see why it is so popular. The most common mode of interaction in computer games is through the keyboard or through game pads. The range of actions that can be symbolically performed in real time through these controls is limited to the physical kind. Players can move the bodies of their avatars, inspect or pick up objects by clicking on them, and trigger the behaviors encoded in these objects, such as firing weapons. The archetypal narrative pattern of the quest makes the most out of these limitations. The hero's deeds are relatively easy to simulate through the game controls, the basic sequence of accomplishment and reward can be repeated endlessly (allowing the player to reach higher and higher levels in the game), the basic script of the quest lends itself to great variations in setting and in the nature of the tasks, and the solitary nature of the hero's quest makes interpersonal relations dispensable. Even when the player needs the assistance of system-created characters or of other players to perform the tasks (as happens in MMORPGs), he advances in the game on his own, and other characters are usually reduced to the fixed roles of either antagonists or helpers. Just as the epic genre highlights the physical deeds of the hero, the games based on this narrative pattern give players the opportunity to distinguish themselves by performing "exploits"—a term which, in hacker jargon, designates an extraordinarily daring and original way to accomplish a task, usually by taking advantage of a glitch in the system. Another reason for the popularity of the epic pattern in video games lies in the graphic capabilities of computers. Epic narratives are basically travel stories that recount the hero's adventures in a world full of danger. The ability of 3D graphic engines to adapt the display to the position of the player's virtual body makes them very efficient at simulating movement as an embodied experience, thereby creating spatial immersion.

Interactivity and the Epistemic Plot

The epistemic plot runs a close second to the epic plot in its compatibility with user interaction. It casts the player in the well-defined role of detective, it combines an authorially defined story—the events being investigated—with a variable story created in real time by the player's actions, it takes advantage of the visual resources of digital systems by sending the player on a search for clues disseminated throughout the storyworld, and it is fully compatible with the types of action that can be easily performed by game controls. These include moving across the world, picking objects, examining them for clues, finding documents, and interrogating nonplaying characters, ideally through a dialogue system but, more efficiently, through a menu of canned questions. An occasional fight against an antagonist can be thrown in to raise the player's adrenaline level.

Interactivity and the Dramatic Plot

The dramatic plot is the most difficult to implement because of its emphasis on the evolution of interpersonal relations. In the goal-oriented action of narrative games based on epic and epistemic plots, NPCs mostly matter to the player because of their capacity to help or hinder the achievement of tasks. Their relation to the player is fixed, and their relations among themselves are practically nonexistent. Very rare are the cases in which the player regards NPCs as human beings rather than as means toward an end. According to Michael Nitsche, such a situation occurs in Deus Ex when the player must kill a formerly friendly NPC who has turned into a zombie in order to progress in the game. In addition, the zombies' pathetic pleas for mercy may soften the player's determination to kill them. Rare in commercial games, this kind of ethical dilemma has become a valued feature in independent games. The moral ambiguity felt by the players—an ambiguity that temporarily distracts them from the game goals—represents a small step from an epic, strictly goal-oriented narrative game to a playable story with a dramatic plot.

The implementation of the dramatic plot raises countless problems. What will be the goals of the player and what kind of algorithm will it take to make these goals interact with the goals of the system-created characters? In an epic plot the player is given a goal by the dispatcher, and all of his efforts are geared toward accomplishing the mission. Similarly, in an epistemic plot, the player remains focused on the elucidation of the mystery until he finds the solution. But in a dramatic plot with evolving interpersonal relations, the characters' goals evolve together with their relations, and they must be constantly redefined. This requires of the system a much more powerful ability to simulate human reasoning than in epic and epistemic plots. Will players spontaneously adapt the goals and plans of their avatars to the current situation, or will the system tell them what to do through a NPC? Will NPCs be equipped with an AI sufficiently sophisticated to read each other's minds, as well as the mind of the player, and to adapt their beliefs, wishes, and plans to every change in situation? The dramatic plot's emphasis on mental states requires from both human and synthetic participants an ability to form representations of the thoughts of other characters, whether this representation is constructed through "theory of mind" or mental simulation. Does she love me or doesn't she? Is he trying to help or to deceive me? Does she intend to keep her promise? Does she know that I know that she loves him? Can I believe what he says, or is he lying? These are the questions that propel and motivate the dramatic plot, whether tragic or comic, the questions that the characters must ask in order to know how to respond to the actions of other characters. The creation of dynamic interpersonal relations between the player and the characters and between the characters themselves makes enormous if not unrealistic demands on the AI that runs the system.[9]

When—and if—all these issues are resolved, the question will remain of what kind of role should be given to the player in order to make the visit to the storyworld into a truly pleasurable experience. I seriously doubt that people would enjoy emulating the heroes of tragedy and comedy by turning themselves into objects of pity or laughter. This suggests that the safest role for the player is that of a marginally involved observer or confidante whose intervention serves mainly as a stimulant that affects the behavior of the synthetic characters and triggers changes in their relations. As a peripheral character the user combines the roles of agent and spectator without assuming responsibility for the story's development and without relinquishing the guidance of an author. As Andrew Glassner observes, most users do not really want to become improvising actors. If they are truly interested in creating their own scripts, they will be much better off participating in online virtual worlds where they will be able to interact with naturally intelligent agents.

Interactivity and the Soap Opera Plot

The open pattern of the soap opera has been successfully adopted by The Sims franchise. As we have already seen, The Sims is a god game in which players create a family of characters and manipulate them from an external/ontological perspective. Players can also adopt characters created by the system who bring a predefined backstory into the game; the player's decisions will continue their lifestory. Each character is defined by individual personality traits, by goals, aspirations, and affective relations to other characters. The game-world also contains a number of NPCs who drop unexpectedly into the life of the characters and interact with them in ways specified by the system. The game thus presents a combination of bottom-up, user-controlled, and top-down, system-controlled storytelling.

The vast number of characters, the types of actions of which they are capable, the possibility for the user to switch control from one character to another, and the fact that the game never ends predisposes The Sims to the generation of multiple episodes and parallel stories driven by love affairs and social relations. Many of the stories created by players can be found on fan fiction Web sites, initially in the form of picture/text combinations (the text being written by the player to complement images generated by the system during gameplay) or, more recently, in the form of videos created by a camera built into the game. Though the external interactivity of The Sims is inherently less immersive than internal participation, the fact that so many people want to share their Sims stories is a powerful testament to the system's immersive potential.

The Pleasure of Interactive Narrative

What kind of reward can we expect from active participation in a story? Narrative pleasure can be generally described in terms of immersion in a fictional world, though some kinds of pleasure lie in distanciation. But a distinction should be made between ludic and narrative immersion. Ludic immersion is a deep absorption in the performance of a task, comparable to the intensity with which a mathematician concentrates on proving a theorem or a soloist performs a concerto. This experience is independent of the mimetic content of the game: players can be deeply immersed in playing chess, go, golf, football, or Tetris—all examples of abstract games—as well as in Deus Ex, Call of Duty, Grand Theft Auto, or the nondigital children's game Cops and Robbers, all games with narrative content. Whereas ludic immersion presupposes a physically active participant, narrative immersion is an engagement of the imagination in the construction and contemplation of a storyworld which relies on purely mental activity. In this section I concentrate on the compatibility of the physically active stance of ludic immersion with the three types of immersion described in chapter 4 as specific to the narrative experience: spatial, temporal and emotional.[10]

Spatial Immersion

Thanks to the visual and animation resources of digital media, spatial immersion is the easiest to achieve in an interactive environment. The ability of digital systems to adapt the display to the position of the player's virtual body makes them very efficient at simulating movement as an embodied experience. Whereas hypertext navigation consists of instantaneous jumps from one node to another that deny the existence of space, the graphic engine of three-dimensional video games and online worlds creates a smooth evolution of the landscape when the user's avatar travels from one point to another. Add to this kinetic experience the visual power of computer graphics, which can make landscapes so beautiful that the exploration of the fictional world becomes an end in itself. The spatial affordances of the medium are most efficiently exploited in epic narratives—stories about the travels of a lonesome hero through a landscape full of dangers—as opposed to dramatic narratives that focus on human relations. The affinity of the epic plot type for character movement is largely responsible for the predominance of the quest pattern in video games.

The user's experience of space in an interactive narrative can take two forms, one strategic and the other emotional. In the strategic experience, space matters to the player for the goal-oriented actions that it makes possible. Many games display space to the player in two ways: (1) as a landscape apprehended from a horizontal point of view, through which the player progresses relatively blindly, since he cannot see the dangers that await him around the next corner, and (2) as a map that represents the world from a vertical point of view. In the vertical projection of maps, no object hides another. By showing resources and obstacles, the map is a precious help to the planning of moves. But the experience of space in a game environment is not only a matter of studying its feature in order to use them to one's advantage, it can also involve an emotional attachment to certain locations. This feeling, which is known to geographers and phenomenologists as "sense of place," is no less compatible with interactivity than the strategic stance. Visitors to the online virtual world Second Life spend lots of time building themselves a personal retreat, a nest lovingly decorated with customized objects made with the tools of the system or bought from other players. While this phenomenon is not strictly narrative, it echoes the importance of the setting for readers of novels. Many people select narratives on the basis of where the action takes place, and according to the cognitive psychologists Jean Mandler and Nancy Johnson ("Remembrance"), setting is the most easily remembered narrative component. Another form of emotional relation to space in digital environments is the kinetic pleasure taken in executing movements. As Gordon Calleja observes, when players achieve a high mastery of the game controls, when movement through game space is a matter of flow and no longer of struggle, they experience the actions of their avatars as if they were performed by their own bodies: "One reason that running, leaping, or flying through a game environment can be so appealing when mastery of the controls has been achieved is that players can then physically and cognitively interpret actions as *their own* running, leaping, and flying, rather than those of an external agent" (*In-Game,* 68, italics original).

Temporal Immersion

Next on the scale of compatibility with interactivity comes temporal immersion, which, as we have seen in chapter 4, Meir Sternberg associates with three narrative effects: curiosity, surprise, and suspense. The first two are relatively unproblematic in interactive environments. Curiosity, or desire to know, is the type of immersion inspired by the epistemic plot. While curiosity can be sustained throughout a narrative, surprise comes in short bursts, inspired by specific events. In an interactive system, the creation of surprise presupposes a top-down generation of events by the system; if the user could fully determine the development of the story, she would not be surprised by the turn of events. A bottom-up system like The Sims can create surprise by temporarily taking control away from the user and generating pre-scripted events, such as the abduction scenario described above. Suspense also requires a top-down management of the player's expectations. People experience suspense when they can foresee two or more possible developments and can't wait to find out which one of these paths will be actualized. The heroine tied to the railroad tracks is a classic instance of suspense because the future can be reduced to an either/or choice: either she will die, or she will be rescued. But when players can determine the plot through their choice of actions, the uncertainty is lost. For suspense to arise in an interactive narrative, the player must *know* of a possible obstacle to his plan or of a counterplan by another character. Take the example of the heroine tied to the railroad tracks: you are playing cowboy Bill, and you want to free her, but you see the train approaching. You experience suspense because you are working in a time frame whose limitations create an obstacle to your goal. There cannot therefore be a more literally temporal kind of immersion. It could be argued that cowboy Bill is too busy saving the heroine to reflect on the situation and that suspense can only be experienced by an external observer. So imagine that you are playing Jack, the villain, and that you need the heroine to die because she saw you rob a bank. You tie her to the railroad tracks, and you gleefully watch the train approach. But suddenly cowboy Bill comes out of nowhere. You first feel surprise, but this surprise turns into suspense when you realize that there is enough time left before the arrival of the train for Bill to rescue the damsel. In this case, your experience will be very close to that of a "passive" reader or a film spectator, since it is created by events that you cannot control, though they crucially affect your goals.

Emotional Immersion

The combination of emotional immersion with interactivity is the most problematic of all the types discussed here because it involves interpersonal relations between the player and computer-operated characters. In real life we experience two main types of emotions—those directed toward ourselves and those directed toward other creatures through a vicarious experience known as empathy. Self-directed emotions concern our desires and the success of the actions through which we try to fulfill them. Even when these emotions involve feelings toward others, such as love and jealousy, the other is an object in a two-way relation determined by the desires of the experiencer. Not so with empathy: it is by mentally simulating the situation of others, by pretending to be them and imagining their desires as our own that we feel joy, pity, or sadness for them.

Narrative has a unique power to generate emotions directed toward others. Aristotle paid tribute to this ability when he described the effect of tragedy as purification (*catharsis*) through feelings of terror and pity inspired by the fate of characters. By contrast, the emotions we experience while playing games—excitement, triumph, dejection, relief, frustration, relaxation, curiosity, and amusement (Lazzaro, "Why We Play Games")—are overwhelmingly self-directed ones because they reflect our success and interest in playing the game. But their range is much smaller than the self-centered emotions of life: computer game players may fight to rescue a princess, and they may receive her hand in reward, but unlike the heroes of love stories, they are not motivated to act by romantic feelings. Whereas narrative interest regards characters as persons, ludic interest regards them as means to an end. Exceedingly rare are the computer-controlled characters who serve not only a functional role by helping or hindering players in the pursuit of their goals but generate interest and empathy through their own personality. I have already mentioned the situation from Deus Ex, in which the player must kill a formerly friendly game-controlled character who has turned into a zombie in order to progress in the game. Some players develop such attachment to this character that they experience extreme discomfort at treating her like an object that needs to be eliminated.

While narrative games deliberately sacrifice characters to action, playable stories have been able to create characters sufficiently lifelike to generate emotional reactions only by limiting the player's participation and hindering self-centered feelings. A case in point is Façade. The interactor experiences intense feelings of dislike and contempt for Grace and Trip, and these feelings are an important part of the interactor's immersion and aesthetic appreciation, but they have no bearing on the development of the story. While the user's agency allows variations in the dialogue that expose diverse facets of Grace and Trip's personalities, the drama unfolds according to a relatively fixed script imposed top-down by the system. The interactor may hate or despise Grace and Trip, but unlike the player of a competitive game or a participant in an online world, she does not entertain strong feelings for her own character, such as caring for her avatar's personal relationship to Grace and Trip. I certainly did not experience sadness over a lost friendship when the couple expelled me from their apartment at the end of the evening to sort out their problems between themselves.

With the epic quest structure of most video games, interactive media have mastered what could be the oldest form of narrative (or at least the oldest form of fictional narrative, for gossip must be older), the struggle of the individual against a hostile world. At the other end of the historical spectrum, the networked structure of hypertext, its fragmentation into recombinant units, and its rejection of the linearity inherent to chronology and causality has made a contribution to the postmodern deconstruction of narrative. The classic narrative form that offers the greatest resistance to active user participation is the dramatic narrative, the type of plot that knots together several destinies into a dynamic network of human relations. Some steps in this direction have been taken with games interspersed with filmic clips, with dialogue systems (Façade), and with simulation algorithms (The Sims), but in all of these approaches, the player's involvement remains peripheral. With film clips she relinquishes agency while the plot is being knotted; with existing dialogue systems she participates in a conversation rather than in a plot, or if there is a plot, she is confined to an observer role; with a simulation she holds the strings of the characters like a puppet master, without personally playing a role in the story. The greatest hurdle to overcome if interactive narrative is to combine the self-centered emotions that come from our active engagement in life and games with the other-centered emotions of traditional narrative is the creation of a kind of relation between the user's character and the synthetic agents that allows the user not only to interact verbally with these agents but also to influence their destiny and to feel personally concerned for both her character and for others.

Conclusion

Throughout this book I have regarded the combination of interactivity, immersion, and narrativity as the formula for total art. The union of narrativity and immersivity has been perfected in literature, film, and drama, but these art forms are not interactive. The marriage of interactivity and immersivity is achieved in our relation to the life-world, but life is not a narrative, though it can be a gold mine of narrative materials when we look at it retrospectively. The reason why we do not (yet?) have "total art" is that the pairing of narrativity and interactivity is much more elusive than the other two combinations, and while it is not impossible, it comes at the price of reduced immersivity, interactivity, or narrativity.

Interactive narrative comes in three forms: actual, near-actual and virtual, which means already implemented, to be implemented in a foreseeable future, and (forever?) under development. The actual forms include hypertext as well as computer games, multimodal stories, alternate reality games (ARGs), and location-based narratives. (A major nondigital form is tabletop role-playing games.) I have already discussed at length the loss of narrative immersion that comes at the cost of interactivity in hypertext, and I will not return to it, but I will take a second look at the relations between interactivity, narrativity, and immersion in games, and a first look at these relations in the other three genres.

Computer games are inherently interactive, and as their technological resources have expanded over the years, their reliance on narrative has steadily increased. Game space is no longer an abstract playing field like a chessboard but a concrete world, populated by both artificially and naturally intelligent agents engaged in the pursuit of their personal interests, and the player's actions determine whether or not these agents will reach their goals. Setting, characters, and events—all the basic elements of narrativity—are currently implemented. Characters are now voiced by highly trained actors, their movements made more realistic through body-capture techniques, and, rather than treating nonplaying characters (NPCs) as one-dimensional enemies or helpers, games may present them as mentally complex creatures who inspire emotions. Meanwhile, narrative discourse (i.e., the dynamic presentation of the story) has adopted many of the techniques of film and literary narratives, such as flashbacks, stream of consciousness (Heavy Rain), and alternation between reality and dream sequences (Max Payne). Cinematic cut scenes, during which the player cannot interact, are often used to build up the game's narrative scaffolding. But the fact that it takes cut scenes to develop interest in the plot further demonstrates the conflict between narrativity and interactivity. Moreover, while games have taken interactivity to unprecedented levels in digital culture, their immersivity remains primarily ludic, that is, a matter of intense concentration on problem solving. Game-worlds, especially the online kind, can also generate a strong sense of place that can be associated with what I have called spatial immersion. This sense of place is nourished by the aesthetics of the visual display, the pleasure of exploring, the pride of being citizens of the game-world, and the folklore that relates to its various locations. But most games subordinate narrative to gameplay, and the commercial kinds rarely deviate from the standard scripts of war and hero's journey. Even a game like Heavy Rain, which has been lauded for the complexity of its underlying narrative (it presents about thirteen different endings, all pre-scripted), requires extensive problem-solving and control-manipulating skills that put it beyond the reach of users who just want a good story with easy interaction. The game industry must have decided that this type of user is too rare to bother with: it is only in independent art games (as well as in simulation games like The Sims) that the story has a chance to become the focus of interest. But except for the Sims type, game narratives are pre-scripted rather than created on the fly. They make no attempt to reach the inner layers of what I call in chapter 7 the interactive onion.

Multimodal stories do not aim at the inner layer either; rather, they take advantage of the ability of digital media to encode data of any kind to propose an experience of the storyworld that encompasses multiple sensory and semantic dimensions. Projects such as *Inanimate Alice*, discussed in chapter 7, or the French *Annalena*, by Daniel Bouillot, about a photographer who discovers that the medium best suited to capture the quality of light of the French Riviera is not photography but painting, fill in the gap that the game industry is reluctant to explore by proposing a user-friendly interface that allows effortless exploration of the storyworld. In *Inanimate Alice*, as we have seen, this exploration takes the form of a progression along a linear story; in *Annalena*, by contrast, only the first and last episodes occupy a fixed spot in the discourse sequence. All the other episodes, which represent self-contained experiences, occur in a random order determined by where the user clicks on a photograph that gradually turns into a painting. The immersivity of multimodal stories lies in the narrative itself and in the sensory richness of the storyworld, not in the opportunity to interact with it, which is mostly an external-exploratory way to bring more information to the screen. Since digital media offer convenient tools for combining sound, text, still picture, and moving picture, multimodal stories can be developed by gifted individuals, and in contrast to computer games, which are very expensive to produce, they are not enslaved to the tyranny of the market. I am personally surprised that not more of the texts in the collections of the Electronic Literature Organization chose this narrative route; the aggressively experimental, worldless approach discussed in chapter 7 still seems to be predominant among practitioners of digital literature.

The genre of the Alternate Reality Game was launched in 2001 with The Beast, a game meant to advertise Steven Spielberg's film *AI*. In an ARG, players reconstitute a story (often a murder mystery) that takes place in a fictional world by following clues that are located in the real world or, more frequently, on the Internet. For instance, a Web site may contain a phone number to call; the person who answers the call may indicate a real-world location to visit where a message will be intercepted. In addition, the player may receive e-mails or SMS messages on her cell phone, and she may get clues from live actors positioned in certain locations. By solving riddles and by communicating with other people, players are directed toward ever-new sources of information that allow them to get more and more of the story. The riddles are usually so difficult that people cannot solve them on their own; in order to progress in the game it is necessary to communicate with other players—in other words, to form communities. At the end of the trail, the players have solved a mystery, much in the same way a detective puts together the story of a case. This story is not really told in any of the fragments—it develops mostly in the mind of the players. Jane McGonigal, a leading designer of ARGs, calls this mode of presentation "chaotic storytelling" (*Reality is Broken*, 287). ARGs, like all games, are inherently interactive, but their interactivity does not concern the story itself, as the participants do not play a role in the storyworld and the plot is entirely scripted by the puppet masters (the producing team). Players may, however, influence the plot indirectly—for instance, when they solve the riddles too fast and force the puppet masters to add some episodes to lengthen the game or, conversely, when the riddles are too hard, and the puppet masters have to introduce more clues. One could speak in this case of meta-interactivity. The passion that some players bring to participating in ARGs demonstrates their immersivity, but while the games revolve around a story, this immersion is not properly narrative but ludic and social. The appeal of ARGs does not reside in the story being discovered but in the activity of solving the mystery by collaborating with other players. Whether ARGs are a passing fad or here to stay is an open question.

Another relatively new genre that involves narrative and interactivity, but without letting interactivity penetrate its stories, is location-based narrative. In this kind of narrative, stories related to visible landscape features become accessible only when users are situated in the *presence* of the referent, in contrast to most texts, which describe absent objects. While location-based narrative does not require digital technology (think of the signs that mark historical sites, telling the story of what happened there), they have received a boost from GPS systems, mobile computing, and cell phones. They are now an established part of the phenomenon of augmented reality—the overwriting of real space with digitally accessible information located "in the cloud." As Scott Ruston writes, "By motivating user activity within a space and layering narrative content on top of physical space, location-based narratives can both reveal the layers of human experience that transform an abstract space into a place and contribute new experience shaping the location into a place" ("Location-based Narrative," 318). A prime example of the power of location-based narrative to carve places out of space is the Canadian project [*murmur*], first developed in Toronto but now implemented in a dozen cities. The ambition of [*murmur*] is to capture the genius loci of a city by collecting stories told by ordinary people about various places and by making them available in these very places through the user's mobile phone. The user is given a map showing the locations that have stories attached to them, and she walks through the city, in search of the stories. Interactivity takes the form shown in figure 17: a freely chosen itinerary that takes the user to the site of noninteractive stories. The immersive power of these stories does not differ from the experience of standard written or oral narrative, except that it could be intensified by the presence of the referent: an immersion both emotional and spatial, since by telling about the people who lived there and shaped the space, the stories create a sense of place.

The near-actual forms of interactive narrative are transpositions of narrative content into three-dimensional, 360-degree interactive environments. Technologies such as those developed by Oculus Rift (HMDs sufficiently light and affordable to be used by individual game players) promise to add a new dimension to the game experience—the dimension that I call spatial immersion. But in contrast to the forms of spatial immersion discussed in chapter 4, this experience will not come from the narrative itself (any story that can be filmed can be shown in a 3D interactive environment) but from the technology that presents it to the senses. Immersive technology can be used to enhance the already interactive narratives of games; experiments by Lugrin and others have shown that players of a first-person shooter game such as Unreal Tournament strongly prefer a three-dimensional stereoscopic environment to a two-dimensional screen. VR can also impart an exploratory kind of interactivity to originally "passive" film narratives. Rather than having the camera determine what fragments of space will be visible, users of a surround system will be able to function as their own camera by simply moving, turning around, and looking at the environment from different angles. (This exploratory function is also possible on the 2D screens of games and online worlds, but it is far less immersive on the screen, since it does not involve the whole of the body.) Vosmeer and Schouten describe the experience of VR cinema as follows: "Instead of telling a whole story in a linear succession of scenes that together form a plot, in this type of scene—that we propose to call *screenscape*—the viewer may be invited to look around and explore, and instead of affecting the plot, participate in the surrounding that the scene is set in" ("Interactive Cinema," 145). When VR technology is used with film, it will be necessary to let the user control the temporal flow of the display, so as to have time to explore every setting before the plot moves the characters to another location. This slowing down may be detrimental to the temporal and emotional forms of immersion. It will take stories in which landscape and setting feature prominently (such as *Lord of the Rings* or the 2014 film *Wild,* about a woman hiking the Pacific Crest Trail) to benefit from the addition of internal, exploratory interactivity and the resulting increase in spatial immersion.

While the actual and near-actual forms of interactive narrative limit interactivity to ways of dealing with fixed narrative scripts, the virtual forms aim toward narrative experiences in which the plot is not predetermined but created on the fly as a response to the user's actions. Since 1999, intensive research has been conducted in the AI community to develop a "narrative intelligence" capable of dynamic story generation. Organizations such as TIDSE (Technologies for Interactive Digital Storytelling and Entertainment), now subsumed under ICIDS (International Conference on Interactive Digital Storytelling), IRIS (Integrative Research in Interactive Storytelling), CMN (Computational Models of Narrative), and a division of AAAI (Association for the Advancement of Artificial Intelligence), hold yearly conferences and workshops on this topic. This research has resulted in one completed and publicly available work, Façade, and in a number of systems and projects under development (Chris Crawford's Storytron, Nicolas Szilas's ID Tension and its practical application, Nothing For Dinner; Cavazza et al.'s interactive version of one episode of *Madame Bovary*), but nothing of commercial or even proven entertainment value has been produced so far. The generation of logically coherent and reasonably varied scenarios has turned out to be a much more complicated endeavor than anticipated in 1999, when the first conference on Narrative Intelligence took place. It is not without reason that projects like Façade (about a couple so self-centered and so mad at each other that they ignore the interactor's contributions to the dialogue) or Nothing for Dinner (about a family trying to cope with the father's loss of memory following an accident) feature characters who behave unpredictably and irrationally: dysfunctional characters enable the system to hide the deficiencies of its logic.

Over the years, researchers have come to realize that if interactive narrative is going to provide an aesthetic experience, its creation should not be entirely left to programmers. While AI people are very good at devising and coding algorithms, it takes creative authors to imagine narrative situations that can incorporate the user's input and produce many interesting variants. The art of interactive storytelling is not a matter of inserting interactive moments within a standard plot but, rather, a matter of discovering what kind of plots lend themselves to active user participation. The focus of research has therefore shifted from the production of interactive narrative to the production of easy-to-use authoring tools. Authors will have to think in terms of situations rich in possible developments, rather than in terms of linear sequence. Ideally, an interactive narrative system will use prewritten elements, such as chunks of dialogue and repertories of actions, in ways that are neither random combinations of hypertext (after two or three transitions, hypertext networks are too complex to be controllable) nor rigidly predetermined sequences. Each narrative action available to the user will have preconditions that can be met only in certain situations and postconditions that generate a new situation. While the author may determine the initial and end state (in order to give closure to the story), there will be many logically coherent ways to connect these two states that the author could not foresee. The difference between a pre-scripted plot that can be implemented in many ways, as shown in figure 20, and a dynamically generated sequence of events may not be qualitative but quantitative. In Façade, for example, there is a global story arc that must be implemented, with a dramatic rise and fall of tension, but the details of the implementation vary with each performance. In Nothing for Dinner, there are plot units that appear in every version (i.e., a female school friend of the son dropping by to get a book and causing the brain-damaged father to behave in an embarrassing way), but the surrounding events vary.

Another recent development is asking why we want interactive narrative rather than assuming that adding interactivity to narrativity will automatically lead to greater immersivity. In the 1990s, an experiment in interactive drama conducted at Carnegie Mellon University as part of the Oz project (directed by Joseph Bates) suggested that the thrill of actively participating in a dramatic scenario would make up for potential losses of narrative interest. In the project (which involved naturally intelligent human actors, not AI-driven characters), an interactor who sees a blind man threatened at knifepoint by a punk at a bus station had to decide whether to shoot the punk, to fire in the air to get the punk to drop the knife, or to "warily edge past the punk" and leave—a rather anticlimactic action, at least from the spectators' point of view. It was found that the external observers were much more critical of the plots than the interactor: "During the experience, the interactors were caught up in the story, did not notice many inconsistencies in either the characters or story, and liked the surprises. In contrast, the observers did not find the characters believable and often lost interest when action seemed to lag" (Kelso et al., "Dramatic Presence," 9). Since the interactor is the real beneficiary of the production, the disparity between the internal and external perspectives represented good news for the future of interactive drama. The active interactor's membership in the storyworld deprives her of the distance necessary to critical judgment. It may be structurally much more difficult to write interactive plots than regular dramatic narratives, but the Oz experiment suggested that this difficulty can be compensated by the user's greater propensity to suspend disbelief in an interactive and physically immersive environment.

More recently, however, researchers have begun to question the intrinsic immersivity of active participation and to wonder what exactly interactive storytelling is trying to accomplish. It is not enough to take a moderately or even intensively interactive narrative and to put it into a VR environment to create the experience of immersive interactivity. The immersion must come from the story, or more precisely, from the user's agency in the storyworld. No longer taking it for granted that more interactivity automatically means greater immersivity and satisfaction, researchers on interactive narrative are more and more interested in collecting user evaluations and in using them as a basis for future work (Milam, El-Nasr, and Wakkary; Friess; Bouillot; Szilas's plans for future development of Nothing

For Dinner). Their goal is to reach an audience that is neither the mass audience of computer games, especially shooters, nor the rarefied audience of the kind of works that are gathered in the two ELO collections. As Michael Mateas, coauthor of Façade, explains, "We are interested in interactive experiences that appeal to the adult, non-computer-geek, movie-and-theater-going public" (Mateas, "A Preliminary Poetics," 29). But this educated public already has novels, movies, comics, and plays to satisfy its need for good stories. Will interactive narrative increase the pleasure taken in traditional media? Will it really be total art? Right now we can only speculate on the basis of the fifteen to twenty minute experiences of Façade or Nothing For Dinner. In the current state of development, the user's interest in interactive narratives with dynamic event generation seems more algorithmic than narrative; by this I mean that rather than experiencing the properly narrative types of immersion—spatial, temporal, or emotional—users will take a metastance, reverse-engineering the code and admiring (or criticizing) the intelligence of the system. Just as Oulipo members viewed literature as a game of overcoming self-imposed constraints (such as writing an entire novel without using the letter *e*, as Georges Perec did in *La Disparition*), we may approach the creation of interactive narratives as a way to solve a particularly tough artistic problem. By creating a distance from the storyworld, such a stance is incompatible with immersion, though it is very compatible with aesthetic appreciation.

If immersive-interactive narrative is a mountain to climb, we have gone a long way toward the top in the past thirty years. We have investigated and largely figured out such topics as how to vary the discourse that reports the story, how to create immersive-interactive visual displays, how to manipulate the camera, how to make characters move more naturally, how to integrate music, and so on. Yet the hardest part of the climb, the dynamic generation of plot in an highly interactive environment, is yet to come—the mountain is a pointed cone like Mount Hood, not a rounded dome like Mount Rainier. But why do we want to climb the mountain so badly? Fanatic alpinists would answer, "Because it is there." Chris Crawford, who has probably devoted more time, energy, and imagination to interactive narrative than anybody else, fully shares this spirit: "To dismiss interactive storytelling on the grounds that it hasn't been done before is to reject the entire basis of the human intellectual adventure" (*Chris Crawford on Interactive Storytelling*, 50). But our fascination for the mountain should not blind us to the surrounding landscape. If we look around, we may discover other peaks in the distance, not quite as steep, that could offer a pleasurable artistic experience because they can be climbed through a viable compromise of immersion, interactivity, and narrativity, a compromise that may lower some of these dimensions in order to enhance the others. In the meantime, we can still dream of an art experience that is not only immersive and interactive, like our relation to the life-world, but that, in contrast to the randomness of life, also offers the design and meaningfulness of narrative. This unique combination, whose formula still eludes us, is what makes it total art.

Notes

CHAPTER ONE: The Two (and Thousand) Faces of the Virtual

Epigraph: Emily Dickinson, "I Dwell in Possibility," in *The Poems of Emily Dickinson.*

1. This essay, which appears on pp. 19–27 of *Art and Artefact* (a tribute to Baudrillard edited by Nicholas Zurbrugg), is synthetically put together out of various passages from *Le Crime parfait.* The translation is not the same as the one that appears in *The Perfect Crime,* the English translation of *Crime parfait* cited in this book.

2. The relation between these two pairs, as well as between the two components of each pair, leaves quite a few unresolved questions. Is the possible opposed to the real, or does Lévy regard the real as a subset of the possible (as does modal logic: the real world is a member of the set of all possible worlds)? Does the actual coincide with the real, or does the real comprise both the virtual and the actual? One solution would be to regard the real as a subset of the possible and the actual and virtual as two modes of being within both the real and the possible. If the design of a computer is a real virtuality, by virtue of being a design, and a specific computer made from this design is a real actuality, then the blueprint of a VR installation that provides a *perfect* sense of the presence of a virtual world would be a possible virtuality, and a particular implementation of this blueprint a possible actuality.

3. Borges has described in "Funes the Memorious" what language would be without this inherent virtuality. Funes, who receives as the result of an accident the dubious gift of total recall, conceives the project of an infinite vocabulary with a different word for every sensory perception: "He was, let us not forget, incapable of ideas of a general, Platonic sort. Not only was it difficult for him to comprehend that the generic symbol dog embraces so many unlike individuals of diverse sizes and form; it bothered him that the dog at three fourteen (seen from the side) should have the same name as the dog at three fifteen (seen from the front)" (*Ficciones,* 114).

4. The weakest of Langer's equivalence, in my view, is the description of poetry as "virtual life" on the ground that a poem creates "a world of its own" (228). Why should poetry be more of a simulation of life than drama and narrative, two genres generally credited with far greater world-creating power than lyric art? The entire discussion of poetry seems symptomatic of the belief, widespread in the era of New Criticism (—when Langer's book was written),—that poetry embodies the essence of language art. As the most sublime of literary genres, it had to virtualize the most "vital" principle, the spark of life itself.

5. See Landow, *Hypertext 2.0,* 7–10, on Bush and the Memex.

CHAPTER TWO: Virtual Reality as Dream and as Technology

Epigraph: Meredith Bricken, "Virtual Worlds: No Interface to Design."

1. Though *virtual reality,* coined by Jaron Lanier in 1989 (Hillis, "Virtual Reality," 512), is the term that has captured the imagination of the general public, arguably because of the poetic appeal of its built-in oxymoron, the scientific community prefers terms such as *artificial reality* (the physico-spatial equivalent of artificial intelligence) or *virtual environments.* The official technical journal of the field, *Presence,* is subtitled *Teleoperators and Virtual Environments.*

2. See Bolter and Grusin, *Remediation,* chap. 16 ("The Virtual Self"), esp. p. 252, for a survey of this controversy.

3. Lanier himself denies deriving any significant inspiration from Gibson and his conception of cyberspace: "In these novels, like . . . *Neuromancer* and so forth, people don't do anything interesting with the artificial reality. . . . Cyberspace is the CB radio of Virtual Reality" (Zhai, *Get Real,* 194).

4. In *The Pearly Gates of Cyberspace,* her outstanding book on cultural conceptions of space, Margaret Wertheim notes (citing Michael Kubovy) that "when we look at a perspectival image from any position other than the center of projection, our minds automatically adjust and we mentally see the image as if we were looking from that point" (114). This phenomenon seems related to the "accommodation to virtual worlds" that helps VR users become immersed in the computer-generated world despite its sensory deficiency with respect to reality.

5. Walter Benjamin observed a similar disappearance of the technological equipment that generates images in the cinema: "Thus, for contemporary man the representation of reality by the film is incomparably more significant than that of the painter, since it offers, precisely because of the thoroughgoing permeation of reality with mechanical equipment, an aspect of reality which is free of all equipment" ("Work of Art," 236). This explains why movies are the most immersive medium to date.

6. Vikram Chandra (*Geek Sublime,* 35) mentions indeed that a player of the game Dwarf Fortress has built a miniature computer within the virtual world of the game.

7. The concept of natural interface, like anything that involves the idea of naturalness, has been under attack by critics who view all behaviors as culturally conditioned. These critics argue that spreading one's fingers to enlarge part of a picture on a touchscreen device is just as conventional as typing a command on a keyboard or using a mouse. But while no mode of interaction with computers is ever fully natural, some are much easier to learn than others because they bear a metaphorical or metonymic relation to the effect to be reached. If it took three taps of the screen to enlarge the picture, this command would have to be learned like an arbitrary code, but spreading one's fingers is more intuitive and easier to remember because it is an analogically motivated gesture. Rather than constituting an inherently natural interface, as some promoters would have us believe (Emerson, *Reading Writing,* 13–19), touchscreen is a semiotic resource out of which more or less motivated signs can be made. Also, since habit is second nature, a switch from keyboard-based to touchscreen interface will be experienced, initially, as unnatural.

8. From an interview published in *Omni* 13 (January 1991).

9. One may, of course, imagine a system doing just that for the sake of aesthetic gratification: an interactive, multimedia implementation of Surrealistic poetry deriving its effect from the incongruity of the metaphor. But in this case the user's action would aim toward magical transformation, not toward sinking golf balls into holes, and the response of the system would fulfill the user's intent.

CHAPTER THREE: The Text as World

Epigraph: Tom Wolfe, "The New Journalism."

1. Thus when Roman Ingarden, founding father of reader-response theory, uses the term *immersion,* he does not mean immersion in a world but immersion in the flow of language itself: "Once we are immersed in the flow of [sentence-thoughts], we are ready, after contemplating the thought of one sentence, to think out the 'continuation'" (quoted in Iser, "Reading Process," 54).

2. As exemplified not only by the theme of the whole book, but by chapter 26 of part 2, where Don Quixote tries to help the heroes of a puppet show by destroying their enemies. He ends up destroying all the puppets and having to pay for them.

3. The standard argument invoked by theorists who oppose the assimilation of fictional

worlds to possible worlds (e.g., Ronen, Walton, Lamarque, and Olsen) is that possible worlds are ontologically complete while fictional worlds are not: the text that describes them cannot specify the truth value of every proposition. But since possible worlds are theoretical constructs, their alleged completeness is not a given but a property ascribed to them by the philosophical imagination. Now, if the imagination can construct possible worlds as ontologically complete, though it cannot run over the list of all propositions, there is no reason it cannot do the same with fictional worlds. The text may produce a necessarily incomplete image, but this does not prevent readers from imagining the textual referents as complete individuals. To take a classic example: We cannot decide how many children Lady Macbeth had, but we imagine her as a woman who had a specific number of children. The number itself is treated as unknown information, not as an ontological gap.

4. The term was first used by Leibniz, but its modern-day users deny any significant indebtedness to Leibniz's philosophy. "It may come as a surprise that this book on possible worlds . . . contains no discussion of the views of Leibniz," writes David Lewis. "Anything I might have to say about Leibniz would be amateurish, undeserving of another's attention, and better left unsaid" (*On the Plurality,* viii).

5. See my article "Possible Worlds in Recent Literary Theory" for an overview of these applications and interpretations.

6. This presentation summarizes ideas that I have presented in *Possible Worlds, Artificial Intelligence, and Narrative Theory.*

7. Philip Zhai has applied the same reasoning to the case of the possible worlds of VR: "We will realize the following symmetry: if we call the actual world real and the virtual world illusory when we are in the actual world, we can also call the actual world illusory and the virtual world real when we are immersed in the virtual world" (*Get Real,* 64).

8. This is not to say that fiction does not involve a moment of evaluation through which the reader distances herself from the image. But this evaluation concerns the art (or performance) of the author, not the truth of the representation. It may involve a comparison with other world-creating performances but not one with the reader's representation of the actual world. This evaluation is therefore not performed from the point of view of a specific world.

9. René Magritte plays with this idea with his painting of a pipe (or more precisely of a shape that evokes the idea of a pipe) containing the words "this is not a pipe."

10. When I speak of universal narrative schema, I do not mean that the ability to produce and understand stories is an autonomous cognitive tool given to us by our biological heritage and that we could locate it in one or more areas of the brain with very advanced imaging (i.e., just as there are Broca's and Wernicke's areas for different aspects of language processing, there would be something we might call "Propp's or Genette's area" exclusively used in storytelling). Rather, I regard this schema as a particular combination of faculties that we employ in other life situations, such as causal reasoning, sequencing, and relating people's actions to their mental states (i.e., mind-reading).

11. Psychologist Marjorie Taylor describes this effect, which was reported by 92 percent of a group of fifty fiction writers, as the "illusion of independent agency" (Keen, "Narrative Empathy," 221).

CHAPTER FOUR: Varieties of Immersion

Epigraphs: Eudora Welty, "Place in Fiction"; Sven Birkerts, *The Gutenberg Elegies;* Marcel Proust, *On Reading.*

1. Balbec is an imaginary name for us but a real one for the narrator of Proust's novel.

2. If we accept David Lewis's definition of fiction as something told as true of a nonactual possible world ("Truth in Fiction"), then the story can be either fictional or nonfictional, depending on its reference world.

3. I do not wish to suggest that this stance is rigidly maintained throughout the "natural" narrative performance. The reduction of spatio-temporal distance at climactic moments through a variety of stylistic devices is a common feature of conversational narration.

4. Chatman is describing here fiction with a third-person impersonal narrator. I show below the phenomenological limitations of this statement. In nonfiction, narrators normally work from memory or documents, but in live broadcasts of sports events, they witness and narrate at the same time. In the fictional domain, the coincidence of seeing and telling is illustrated by narrators who describe movies for the blind or by those African film narrators who give voice to the action on the screen because the film is in a foreign language. In these last two cases, the narrator is a real-life person who does not belong to the fictional world.

5. By discussing the narrative devices that lead to immersion, I do not mean that their use is automatically immersive, nor that they are required for immersion. There are certainly other ways to immerse readers. According to what Sternberg ("Proteus in Quotation-Land") calls "the Proteus principle," narrative devices cannot be ascribed exclusive functions. All my immersive devices could reach the opposite effect if used without subtlety.

6. An opposite, equally important trend consists of multiplying versions of events in an attempt to stress the artificiality of narrative form; what is lost between these two extremes is the belief that narrative can be an authentic expression of life experience.

7. This distinction between logical and imaginative recentering translates into my own model and terminology some concepts worked out by Zubin and Hewitt in their study of deixis in narrative ("Deictic Center").

8. To my knowledge, the only "textualist" attempt to explain the phenomenon of emotional participation is Richard Walsh's essay "Why We Wept for Little Nell: Character and Emotional Involvement." Faithful to the structuralist view that characters are not human beings but textual constructs, Walsh advances the thesis that "the reading of fiction requires evaluation, but not belief or any simulacrum of belief" (312). In other words, the reader is invited to contemplate propositions but not to become immersed in a virtual world. While the Victorian reader's tears for Little Nell can be partly explained by Victorian cultural habits, the modern reader's feelings of sadness for Dickens's heroine are "a response to the idea of innocence . . . rather than to the innocent girl to which that idea contributes. . . . Instead of saying that readers' emotional responses to the fortunes of a character are the result of involvement with a represented person, [this account] assumes that their emotions attach to the particular complex of meanings constituting the character" (ibid.). In this account, the reader would be moved by the abstract idea that innocent little girls can die rather than by the death of one particular fictional individual. Along this line of thought, our evaluations of allegories (such as we find in *The Faerie Queene*) would be just as emotional as our responses to the characters of realistic fiction, or perhaps more so, since allegories embody ideas more clearly than complex lifelike characters do. This sounds highly counterintuitive.

9. James Phelan *(Reading People)* distinguishes three components of character: "synthetic" (character as aggregate of features selected by writer), "mimetic" (character as person), and "thematic" (character as bearer of some sort of significance). The world-centered approach emphasizes the mimetic dimension, but it does not preclude the other two, since it maintains an ontological distinction between real and pseudo persons. In this perspective, readers can either play the game of make-believe and regard characters as persons or adopt a distanced ("metafictional," "textualist") position and regard them as verbal creations fulfilling a thematic function. The textual and world approaches are complementary, but it is only in the world approach that emotional reactions can be justified.

10. Another example of fear that brings into play the thought that a fictional situation might be actualized and affect the spectator is arguably Aristotle's concept of terror. According to his

translator Malcolm Heath (xxxix), this terror brings relief and leads to pleasure because it acquaints the spectator with the feeling of fear and purges him of his tendency to surrender to excessive or inappropriate fear in real life. Through *katharsis,* according to this analysis, tragedy takes the sting out of existential anguish.

CHAPTER FIVE: The Text as World versus the Text as Game

Epigraph: Lewis Carroll, *Through the Looking-Glass*

1. On these differences, see R. Rawdon Wilson, *In Palamedes' Shadow,* chap. 3, esp. 98–99.

2. This process of deciphering the rules has been invoked by Wolfgang Iser in support of the text/game analogy: "[The aesthetic pleasure of the text is heightened] when the text keeps its rules of play hidden, so that their discovery can become a game in itself. . . . [The] pleasure will increase when the rules found challenge the sensory and emotive faculties" (*The Fictive and the Imaginary,* 278).

3. The difficulty of fruitfully applying the notions of winning and losing to literature explains why the mathematical field of game theory and Jaakko Hintikka's game-theoretical semantics have had little impact on literary theory. Both are based on the idea of competition between players.

4. In treating *afternoon* as a game, I am disregarding this note on the package of the text: "Michael Joyce's *afternoon* is a pioneering work of literature, a serious exploration of a new hypertextual medium. It is neither a game nor a puzzle." But why couldn't a "serious" use of hypertext, constituting a "pioneering" literary text, take the form of a game?

5. This concept is proposed in *Struktura khudozhestvennovo teksta* (*Structure of the Literary Text*) (Moscow: Iskutsstvo, 1970). See the entry for Lotman by Eva Le Grand in Makaryk, *Encyclopedia,* 407–10.

6. Cf. Derrida's famous phrase "Il n'y a pas de hors-texte" (There is nothing outside the text).

7. As in Aya Karpinska's digital poem "The Arrival of the Bee Box."

8. The translation by Richard Howard has "functioning" here, but "playing" is closer to the French, "jouer" ("Ce lecteur est . . . plongé dans une sorte d'oisiveté, d'intransivité, et pour tout dire, de sérieux: au lieu de jouer lui-même, d'accéder pleinement à l'enchantement du signifiant, à la volupté de l'écriture, il ne lui reste plus en partage que la pauvre liberté de recevoir ou de rejeter le texte").

CHAPTER SIX: Texts without Worlds

1. This is not to say that art and play are devoid of real-world functionality; scholars such as Brian Boyd and Denis Dutton have devoted entire books defending their adaptive value, i.e., their contribution to the survival and evolution of mankind.

2. Quoted from www.netbehaviour.org/pipermail/netbehaviour/20080131/007271.html.

3. This is how Nelson describes another of his works, "Scrape Scraperteeth," in an interview with the *Guardian.* See www.theguardian.com/technology/gamesblog/2011/sep/13/the-meaning-of-art-games.

4. See selections from the Lettrist Manifestos on Isidore Isou's home page, www.thing.net/~grist/l&d/lettrist/isou-m.htm.

5. See Simanowski's *Digital Art and Meaning* (35–42) for a lengthy discussion of whether or not *Text Rain* qualifies as literature.

6. This conversation was sparked, to a large extent, by Franco Moretti's practice of "distant reading" (i.e., relying on other people's readings, on summaries, and on automated word searches) to study large corpuses of literary texts. This type of reading challenged the sacrosanct "close reading" that is traditionally practiced in academia.

7. It is tempting to associate deep attention with immersion and with the print medium,

and hyper attention with interactivity and with digital media, but there are many exceptions to this rule. A Dickens novel read on a computer screen, for a reader familiar with the application, is fully capable of generating immersive deep attention, and so does a good video game, while a print newspaper invites readers to move back-and-forth between titles, short texts, and images, through a process not all that different from surfing the Web. It is the structure of a text, much more than the medium (if we understand medium as supporting technology), that determines the mode of reading and the quality of attention.

8. The only project discussed in this chapter whose formula would be worth repeating is the computer language Shakespeare because the generative formula is very difficult to implement, and it can consequently be implemented with greater or lesser success. If the formula is repeatable, it is more a formal constraint—similar to the constraints that define the genre sonnet—than a generative idea.

CHAPTER SEVEN: The Many Forms of Interactivity

1. In *Orality and Literacy* (136), Walter Ong argues indeed that electronic media are bringing an age of "secondary orality" characterized by a participatory mystique, communal sense, concentration on the present moment, and use of formula. Writing in 1982, Ong could not have had the Internet in mind (he was thinking mostly of radio and TV), but each of the three points mentioned above can be illustrated by digital phenomena: communal sense by social media, concentration on the present moment by real-time chat, and use of formulae by emoticons and acronyms such as LOL.

2. Jane Yellowlees Douglas captured this preservation of openness by describing the forking paths of hypertext as implementing a relation of AND / AND / AND rather than the exclusive OR.

3. Mark Bernstein has proposed a typology of hypertext patterns that could be used to create further refinements within the network category. His catalog of patterns includes cycle, counterpoint, mirror worlds, tangle, sieve, montage, split/join, missing link, and feint. Split/join corresponds to what I describe as the flowchart, and the sieve corresponds to the tree design. But not all of these patterns concern the basic shape of the network. The difference between counterpoint and mirror world is mainly thematic (different voices focused on different themes versus different voices presenting the same themes from different points of view), montage and feint are styles of visual presentation that seem compatible with several network configurations, and missing link is a matter of unfulfilled expectations. It is therefore impossible to subsume Bernstein's typology within the one I am presenting here, nor vice versa.

4. An example of this strategy is a combinatory play written by the French Oulipo members Pierre Fournel and Jean-Pierre Enard (Motte, *Oulipo: A Primer of Potential Literature,* 156–58). By allowing crossover and closing the tree at the bottom, they were able to write a system that generates sixteen plays with only fifteen different "scenes," some of which allow no choice. They claim that their scheme saved them sixty-seven scenes, a substantial reduction of memorization for the actors, but a simple calculation shows that a binary tree of five levels with four decision points requires only thirty-one scenes (16+8+4+2+1)! Fournel and Enard assume that sixteen plays of five scenes would take 16×5=80 scenes, but the sixteen plays would not form a combinatorial system, since they would be totally independent of each other.

5. Janet Murray (*Hamlet on the Holodeck,* 158–59) describes a play that implements a closely related idea, Alan Ayckbourne's *The Norman Conquest,* which consists of three different plays taking place simultaneously in different rooms of the same house, but the production does not seem to allow the spectator to move from room to room. The plays are therefore parallel but not interactive.

6. Here is an example of quest in World of Warcraft: "The ancient prophecy of Mosh'aru speak of a way to contain the god Hakkar's essence. It was written on two tablets and taken to the

troll city of Zul'farrak, west of Gadgetstan. Bring me the Mosh'aru tablets. The first tablet is held by the long dead troll Theka the Martyr. It is said his persecutors were cursed into scarabs and now scuttle from his shrine. The second is held by the hydromancer Vlratha, near the sacred pool of Gahz'rilla. When you have the tablets, bring them to me" (quoted from Krzywinska, "World Creation and Lore," 129).

7. If the game were not replayable, users would be unable to find out if the program really listens to their input. There are indeed some examples of fake interactivity, where the system asks questions to the user but disregards the answers and tells the same story every time.

8. WYSIWYG stands for What You See Is What You Get, the slogan of Graphic User Interface.

CHAPTER EIGHT: Hypertext

Epigraphs: Jorge Luis Borges, "An Examination of the Work of Herbert Quain," in *Ficciones;* Walter Benjamin, "The Work of Art in the Age of Mechanical Reproduction."

1. Hypertext as a concept is much older. Ted Nelson introduced it in 1963. See the Wikipedia entry "Hypertext" (http://en.wikipedia.org/wiki/Hypertext) for the history of the term.

2. The chapter is titled "The Extraordinary Convergence: Democracy, Technology, Theory, and the University Curriculum."

3. In her later essay, "Beyond the Hype: Reassessing Hypertext," Snyder takes a much more skeptical approach to the claim that hypertext represents the fulfillment of postmodern doctrine.

4. While I agree in its broad lines with Manovich's opposition, I have problems with his formulation, particularly with the phrase "seemingly unordered." To whom do the events seem to lack order? Not to the author, who imagines event sequences, nor to the reader, who reconstructs these sequences out of the text's information.

5. The most database-like hypertext that I know is Deena Larsen's *Marble Springs,* discussed in chapter 7. In the case of *Marble Springs* the database structure does not impede narrativity because the texts consist of small stories contained in individual lexias. Since these stories generally involve multiple places and characters, they are accessible in many different ways.

6. It could be argued that if hypertext generates a new story with every visit, clicking would indeed have ontological consequences (Bode, *Future Narratives,* 53). However, in this chapter, I dispute this view.

7. In *afternoon,* for instance, Joyce uses guard fields to ensure that the reader cannot reach the screen that suggests Peter's possible responsibility for the accident that (perhaps) killed his son and ex-wife before Peter and the reader have gone through a therapy session with the psychologist Lolly. This reading is developed by J. Yellowlees Douglas in "How Do I Stop This Thing."

8. This is Brian McHale's term (in chapter 7 of *Postmodernist Fiction*) for the postmodern practice of creating and destroying fictional worlds.

9. See Mark Nunes, "Smooth and Striated Cyberspace," for an enlightening application of these concepts to electronic culture.

10. Could there be more than one image? An example from print literature suggests that this could be the case. In Robert Coover's "The Babysitter" (in *Pricksongs and Descants*), the reader encounters a series of short paragraphs that describe an evening spent by a teenager babysitting two kids. But it quickly turns out that the various fragments do not tell the same story: some form a sequence in which the teenager arranges a visit by her boyfriend at the children's home while the parents are away, one in which she is murdered by two intruders, one in which everything unfolds normally. There are consequently distinct possible scenarios. As readers progress sequentially through the text, they sort out the narrative material and build several alternative possible worlds by assigning segments to the proper sequence. (A given segment may fit into more than one narrative script.) The text builds the various stories in a loose round-robin fashion, adding to one, then to another, but without disrupting chronological sequence.

11. See Bell, *The Possible Worlds of Hypertext Fiction*, for a systematic treatment of this approach.

12. William Gibson, the coiner of the term, calls it a "nonspace of the mind" (*Neuromancer*, 57).

13. Far from being satisfied with her own diagnosis, Murray tries to make a case for the tragic and cathartic potential of electronic narrative by imagining three interactive ways of representing the journey of a young man toward suicide (175–82). But while a gifted writer could conceivably manage to create emotional bonding of the reader with the character, this accomplishment would be more a matter of overcoming the limitations of the medium than of exploiting its distinctive properties.

CHAPTER NINE: Participatory Interactivity from Life Situations to Drama

Epigraph: Susanne Langer, *Feeling and Form.*

1. I do not wish to convey the impression that VR developers and theorists regard the mutual compatibility of immersion and interactivity as totally unproblematic. Jonathan Steuer suggests that a very immersive VR display may "decrease the ability of subjects to mindfully interact with it in real time" ("Defining," 90). If a computer-generated environment is so rich in "fictional truths" that its exploration offers great rewards, why would the user bother to change this world? It is, however, through exploration that the user gains an appreciation for the immersive richness of the virtual world, and this exploration is itself a type of interactivity. Moreover, the problem suggested by Steuer would be a case of immersion interfering with interactivity—the familiar objection that immersion makes the user passive—while I am debating here the opposite question: Does interactivity promote immersion in three-dimensional VR environments?

2. Through this comparison I do not wish to deny that authors use materials from their own life experience. In the domain of art, literally speaking, there is no creation ex nihilo.

3. See Angela Ndalianis, *Neo-Baroque Aesthetics and Contemporary Entertainment*, on the affinities between Baroque art and digital culture.

4. Painted in 1751 by Franz Josef Spiegler.

5. See Omar Calabrese, *Neo-Baroque: A Sign of the Times*, for a discussion of the affinities between Baroque art and postmodernism. The chapter titles tell much of the story: "Rhythm and Repetition," "Limit and Excess," "Detail and Fragment," "Instability and Metamorphosis," "Disorder and Chaos," "The Knot and the Labyrinth," "Complexity and Dissipation," "The Approximate and the Inexpressible," "Distortion and Perversion." In *Neo-Baroque Aesthetics and Contemporary Entertainment*, Angela Ndalianis extends the parallelism to the visual culture of the digital age by stressing such Baroque features as illusionism, sensorial excess, optical virtuosity, and dynamic shapes conducive to mental or physical movement in video games and special effect movies.

6. This was Nietzsche's objection to the concept: "Now the serious events are supposed to prompt pity and fear to discharge themselves in a way that relieves us; now we are supposed to feel elated and inspired by the triumph of good and noble principles, at the sacrifice of the hero in the interest of a moral vision of the universe. I am sure that for countless men this, and only this is the effect of tragedy, but it plainly follows that all of these men, together with their interpreting aestheticians, have had no experience of tragedy as a supreme art" (*Birth of Tragedy*, sec. 22, p. 32).

7. As a counterexample to this principle, one should mention that privileged noblemen had a seat on stage in Parisian theaters of the time of Louis XIV, but this practice was a leftover from the Elizabethan age and would later disappear.

8. As Barthes puts it, "Why is the writerly our value? Because the goal of literary work (of literature as work) is to make the reader no longer a consumer, but a producer of the text" (*S/Z*, 4).

9. Hillis writes, "In [*The Theater and its Double*] Artaud makes the link between alchemical practice and the theater, asserting that theater can create a virtual reality (*la réalité virtuelle*) that transforms matter from spirit or mind and in which characters, images and objects 'take on the

phantasmagoric force of alchemy's visionary internal drama'" ("Virtual Reality," 510; Hillis quotes Erik Davis, *Myth, Magic and Mysticism in the Age of Information*).

CHAPTER TEN: Chasing the Dream of the Immersive, Interactive Narrative

1. I do indeed conceive narrativity as a scalar property, fully realized in genres such as classical drama and comedy, epics, anecdotes, narratives of personal experience, TV serials and series, short stories, and, until the mid-twentieth century, novels, but only partially realized in experimental novels, installation art, and in many of the interactions of everyday life. See Ryan (*Avatars*) on my conception of the conditions of narrativity.

2. Archeologists are still unsure of the precise social function of the kivas.

3. On this interaction between users, Katherine Hayles writes, "Each can see the other in the simulation and hear the voice-filtered comments his or her companion makes. Participants can shapeshift by touching the appropriate totemic icon. Improvising on cues provided by the environment and each other, they create narrative" ("Embodied Virtuality," 19).

4. Only Snake's vision was implemented and, according to the authors, with limited success: "The implementation was poor, in that it simply applied a red filter without increasing apparent luminance, thus effectively reducing rather than enhancing visibility" (123).

5. I adapt this term of "playable stories" from Pat Harrigan and Noah Wardrip-Fruin's concept of "playable media" (*Second Person*).

6. My categories of playable story and narrative game correspond to some extent to the categories of ludic digital literature and literary computer games proposed by Astrid Ensslin (*Literary Gaming*, chapter 3), but I avoid the term *literature* because nowadays most digital narratives are multimodal, and I do not want to restrict my inquiry to works in which language is dominant. Also, while Ensslin's literary games are characterized by an artistic quality, my narrative games can be both independent art games and commercial games produced by large companies, as long as they enact a story.

7. A computer passes the Turing test (devised by British mathematician Alan Turing) when a human judge interacting verbally with a computer and a human being cannot tell who is who. Success in the Turing test is measured in terms of how long it takes for the judge to make the correct identification. The Loebner prize requires contestants to last five minutes.

8. With the development of the novel and even earlier in *The Iliad*, the prototypical epic and dramatic plot were hybridized into open plots and action-centered stories that present much more complex interpersonal relations than the archetypal fairy tale.

9. These demands can be reduced in a system that generates variations on a known plot, such as the interactive version of *Madame Bovary* developed by Cavazza et al. ("*Madame Bovary* on the Holodeck"). In this kind of system, the user learns in advance about the characters, their personal goals and the general situation, and he impersonates one of the characters. This limits the range of possible actions and the number of rules required by the system. In Cavazza's implementation, the user plays Rodolphe, Emma's lover, and he can respond in various ways when Emma suggests that they elope together. The system, however, is limited to one scene, which could end in various ways, but some of these ways would most likely be incompatible with the plot of the novel as a whole. If the interactive system were extended beyond a single scene, it would have to bring back the plot in a proper direction (i.e., one reasonably faithful to the novel); in other words, if Rodolphe accepts Emma's suggestion, then the system would have to create some catastrophic event that would force Emma to return to her husband so that she can fulfill her tragic destiny.

10. Another kind of immersion, provided by certain types of games, that is not particularly narrative is social immersion. Multiplayer online worlds are as much forums for meeting other players as they are arenas for displaying skills and progressing in the game hierarchy. Some of the tasks presented to the player can only be performed by forming alliances and sharing resources

with other players. These alliances, known as guilds, lead to a strong sense of belonging to a virtual community. See Calleja, *In-Game,* for a taxonomy of the various types of immersion (or engagement) provided by computer games. His system comprises kinesthetic involvement (part of my spatial immersion), spatial involvement, shared involvement (my social), narrative involvement (which I further subdivide), affective involvement (my emotional), and ludic involvement.

Works Cited

PRIMARY, PRINTED

Balzac, Honoré de. *Eugénie Grandet.* Trans. Marion Ayton Crawford. London: Penguin, 1955.

Barnes, Julian. *Flaubert's Parrot.* New York: Knopf, 1985.

Boccaccio, Giovanni. *The Decameron.* Trans. and ed. G. H. McWilliam. New York: Penguin, 1972.

Borges, Jorge Luis. *Ficciones.* Ed. and intro. Anthony Kerrigan. New York: Grove Weidenfeld, 1962.

Brontë, Charlotte. *Shirley.* Oxford: Clarendon, 1979.

Brontë, Emily. *Wuthering Heights.* New York: Bantam, 1983.

Carroll, Lewis. *Alice's Adventures in Wonderland and Through the Looking Glass.* London: Puffin Books, 1948.

Conrad, Joseph. *The Nigger of the Narcissus.* London: Dent, 1974.

Coover, Robert. *Pricksongs and Descants.* New York: New American Library, 1969.

de Cervantes Saavedra, Miguel. *Don Quixote of La Mancha.* Trans. Walter Starkie. New York: Signet, 1994.

Dickinson, Emily. *The Poems of Emily Dickinson.* Variorum edition. Ed. R. W. Franklin. Cambridge, MA: Belknap, 1998.

Duras, Marguerite. *L'Amant.* Paris: Minuit, 1984.

Flaubert, Gustave. *Madame Bovary.* Trans. Francis Steegmuller. New York: Random House, 1957.

Fowles, John. *The French Lieutenant's Woman.* New York: Signet, 1970.

Gibson, William. *Neuromancer.* New York: Ace, 1994.

Huxley, Aldous. *Brave New World.* New York: Harper Perennial, 1989.

Joyce, James. "Eveline." *Dubliners.* New York: Penguin, 1976.

Lightman, Alan. *Einstein's Dreams.* New York: Vintage, 2004.

Loyola, Ignatius of. *The Spiritual Exercises and Selected Works.* Ed. George E. Ganss, S.J. New York: Paulist Press, 1991.

Morrison, Toni. *Beloved.* New York: Plume/Penguin, 1988.

Pavić, Milorad. *Dictionary of the Khazars: A Lexicon Novel in 100,000 Words.* Trans. Christina Pribicevic-Zoric. New York: Knopf, 1988.

Perec, Georges. *La Disparition.* Paris: Denoël, 1969.

———. *La Vie mode d'emploi.* Paris: Hachette, 1978.

Pessl, Marisha. *Night Film.* New York: Random House, 2013.

Poe, Edgar Allan. "The Assignation." *The Portable Poe.* Ed. Philip Van Doren Stern. New York: Penguin-Viking, 1997.

Proust, Marcel. *On Reading.* Trans. and ed. Jean Autret and William Burford. Intro. William Burford. New York: Macmillan, 1971.

———. *Remembrance of Things Past.* Vol. 1, *Swann's Way, Within a Budding Grove, The Guermantes Way.* Trans. C. K. Scott Moncrief. New York: Random House, 1934.

Queneau, Raymond. *Cent mille milliards de poèmes.* Paris: Gallimard, 1961.

Robbe-Grillet, Alain. *Two Novels by Robbe-Grillet: Jealousy and In the Labyrinth.* Trans. Richard Howard. New York: Grove, 1965.

Roubaud, Jacques. ∈. Paris: Gallimard, 1967.
Saporta, Marc. *Composition No 1*. Paris: Seuil, 1961.
Sebald, W. G. *The Emigrants*. Trans. Michael Hulse. New York: New Directions, 1996.
Stendhal [Henri Beyle]. *Le Rouge et le noir: Chronique du XIXème siècle*. Ed. Henri Martineau. Paris: Garnier, 1958.

PRIMARY, DIGITAL

Amerika, Mark. *Grammatron*. www.grammatron.com.
Ankerson, Ingrid, and Megan Sapnar. "Cruising." 2001. www.poemsthatgo.com/gallery/spring2001/crusing-launch.html.
Armstrong, Kate, and Michael Tippett. *Grafik Dynamo*. 2005. www.turbulence.org/Works/dynamo/.
Beiguelman, Giselle. "Code Movie 1." 2004. *Electronic Literature Collection* 1. http://collection.eliterature.org/1/works/beiguelman__code_movie_1.html.
Bouilllot, Daniel. Annalena. Multimodal narrative, privately distributed. Description at www.lisiere.com/annalena/index.htm.
Breeze, Mez. "[NetBehaviour] _531chambered:amber:hammered:heart:Kernel_." www.netbehaviour.org/pipermail/netbehaviour/20080131/007271.html.
Brown, Marc. *Arthur's Teacher Troubles*. CD ROM. Brøderbund, 1996.
Cayley, John. "The Speaking Clock." 1995. Original version does not run on today's computers. A new version can be found at http://programmatology.shadoof.net/index.php?p=works/clocks/clocks.html.
———. "Translation." 2004. *Electronic Literature Collection* 1. http://collection.eliterature.org/1/works/cayley__translation.html.
Civilization. Video game. Designed by Sid Meier and Bruce Shelley, published by Microprose and Koei, 1991–2010.
Coverley, M. D. *Califia*. Cambridge, MA: Eastgate, 2001.
Deus Ex. Video game. Developed by Ion Storm, published by Eidos Interactive. 2000–2012.
Electronic Literature Collection, vol. 1 http://collection.eliterature.org/1/.
Electronic Literature Collection, vol. 2. http://collection.eliterature.org/2/.
Falco, Edward. *Chemical Landscapes*. 2006. Electronic Literature Collection 1. http://collection.eliterature.org/1/works/falco__chemical_landscapes_digital_tales.html.
geniwate, and Deena Larsen. *The Princess Murderer*. 2003. www.deenalarsen.net/princess/.
Glaser, Michelle, Andrew Hutchison, and Marie-Louise Xavier. *Juvenate: An Interactive Narrative*. CD ROM. Produced in association with the Australian Film Commission.
Heavy Rain. Video game. Directed by David Cage. Quantic Dream for Sony Computer Entertainment, 2011.
I'm Your Man. Dir. Bob Bejean. Perf. Mark Metcalf, Colleen Quinn, and Kevin M. Seal. A Choice Point Film. Presented by Planet Theory in association with DVD International. DVD edition produced by Bill Franzblau, 1998.
Joyce, Michael. *afternoon, a story*. Hypertext software. Cambridge, MA: Eastgate Systems, 1987.
———. *Twelve Blue: Story in Eight Bars*. World Wide Web hyperfiction. *Postmodern Culture* and Eastgate Systems, 1996 and 1997. www.eastgate.com/TwelveBlue/.
Karpinska, Aya. "Arrival of the Bee Box." www.technekai.com/box/.
Larsen, Deena. *Marble Springs* 3.0. http://marblesprings.wikidot.com/start.
Laurel, Brenda, Rachel Strickland, and Rob Tow. *Placeholder*. VR installation. Produced by Interval Research Corporation and the Banff Centre, 1993. Description at www.interval.com/frameset.cgi?projects/placeholder/index.html.

The Manhole. Video game. Designed by Rand and Robyn Miller, published by Brøderbund, 1988–2007.

Mateas, Michael, and Andrew Stern. Façade. 2005. Downloadable from http://interactivestory.net/download.

Max Payne. Computer game. Developed by Remedy Entertainment. Rockstar Games, 2003.

Morrissey, Judd, and Lori Talley. *The Jew's Daughter.* 2000. *Electronic Literature Collection* 1. http://collection.eliterature.org/1/works/morrissey__the_jews_daughter.html.

Moulthrop, Stuart. *Hegirascope.* World Wide Web hyperfiction. http://raven.ubalt.edu/staff/moulthrop/hypertexts/hgs/.

———. *Marginal.* 2005. www.tekka.net/07/marginal/mfx01.swf.

———. *Reagan Library.* 1999. *Electronic Literature Collection* 1. http://collection.eliterature.org/1/works/moulthrop__reagan_library.html.

———. *Victory Garden.* Hypertext software. Cambridge, MA: Eastgate Systems, 1991. [*murmur*]. Locative narrative, Toronto version. http://murmurtoronto.ca/.

Myst. Video game. Directed by Rand and Robyn Miller. Cyan for Red Orb Entertainment, 1993.

Nelson, Jason. "Game, Game, Game and Again Game." 2007. *Electronic Literature Collection* 2. http://collection.eliterature.org/2/works/nelson_game.html.

Nothing for Dinner. Interactive narrative, developed at the University of Geneva. http://tbisim.unige.ch/portal/?q=node/21.

Powers, Richard, with Jenifer Gunji, Joseph Squier, Jessica Mullen, Lauren Hoopes, Chad Kellenberger, and Val Lohmann. "They come in a steady stream now." 2004. www.ninthletter.com/featured_artist/artist/5.

Rosenberg, Jim. "The Barrier Frames" and "Diffractions Through." Software. *Eastgate Quarterly Review of Hypertext.* Cambridge, MA: Eastgate Systems, 1996.

Pullinger, Kate, and Chris Joseph. *Inanimate Alice.* www.inanimatealice.com/.

Schleiner, Anne Marie. "Velvet Strike." Demo at https://www.youtube.com/watch?v=yAhGoPJBQAA.

Second Life. Online world. Developed by Linden Labs, 2003–present.

The Sims. Video game. Designed by Will Wright. Maxis, 2000–2014.

Stefans, Brian Kim. *The Dream Life of Letters.* 2000. *Electronic Literature Collection* 1. http://collection.eliterature.org/1/works/stefans__the_dreamlife_of_letters.html.

Unreal Tournament. Video game. Designed by Cliff Blezinski and James Schmalz, published by GT Interactive, Infogames and Redversion Games, 1999–2001.

Utterback, Camille, and Romy Arhitruv. *Text Rain.* 1999. Description at www.camilleutterback.com/textrain.html.

Ward, Adrian. *Auto-Illustrator.* http://download.cnet.com/Signwave-Auto-Illustrator/3000-2191_4-9580.html.

Wardrip-Fruin, Noah, and Brion Moss, with a. c. chapman and Duane Whitehurst. *The Impermanence Agent.* No longer available, documentation at www.impermanenceagent.org/agent/.

World of Warcraft. Video game. Designed by Rob Pardo, Jeff Kaplan, and Tom Chilton. Blizzard Entertainment, 2004–2011.

SECONDARY, PRINTED AND DIGITAL

Aarseth, Espen. "Aporia and Epiphany in Doom and The Speaking Clock: The Temporality of Ergodic Art." Ryan, *Cyberspace,* 31–41.

———. *Cybertext: Perspectives on Ergodic Literature.* Baltimore: Johns Hopkins UP, 1997.

Abbott, H. Porter. *The Cambridge Companion to Narrative.* 2nd ed. Cambridge: Cambridge UP, 2008.

Aristotle. *Poetics.* Trans. and intro. Malcolm Heath. New York: Penguin, 1996.

Artaud, Antonin. *Artaud on Theater.* Ed. Claude Schumacher. London: Methuen, 1989.

Augé, Marc. *Non-Places: Introduction to an Anthropology of Supermodernity.* New York: Verso, 1995.

Aylett, Ruth, and Sandy Louchard. "The Emergent Narrative: Theoretical Investigation." *Proceedings of the Narrative and Learning Environments Conference,* NILE04, Edinburgh, Scotland, 25–33. 2004. www.nicve.salford.ac.uk/sandy/ENFramesetPage.htm.

Bachelard, Gaston. *The Poetics of Space.* Trans. Maria Jolas. Boston: Beacon, 1994.

Barthes, Roland. "The Death of the Author." *Image, Music, Text.* Ed. and trans. Stephen Heath. New York: Hill & Wang, 1977. 142–48.

———. *Le Plaisir du texte.* Paris: Seuil, 1973.

———. "The Reality Effect." *French Literary Theory Today.* Ed. Tzvetan Todorov. Cambridge: Cambridge UP, 1982. 11–17.

———. *S/Z.* Trans. Richard Howard. New York: Hill & Wang, 1974.

Bates, Joseph. "The Nature of Characters in Interactive Worlds and the Oz Project." Loeffler and Anderson, *Virtual Reality Casebook,* 99–102.

———. "The Role of Emotions in Believable Agents." *Communications of the ACM* 37, no. 7 (1994): 122–25.

Baudrillard, Jean. *Art and Artefact.* Ed. Nicholas Zurbrugg. London: Sage, 1997.

———. *The Perfect Crime.* Trans. Chris Turner. London: Verso, 1996.

———. "The Precession of Simulacra." *Simulacra and Simulation.* Trans. Sheila Faria Glaser. Ann Arbor: U of Michigan P, 1994. 1–42.

Bell, Alice. *The Possible Worlds of Hypertext Fiction.* Chippenham: Palgrave MacMillan, 2010.

Benedikt, Michael. "Introduction" and "Cyberspace: Some Proposals." Benedikt, *Cyberspace,* 1–16, and 119–224.

———, ed. *Cyberspace: First Steps.* Cambridge, MA: MIT Press, 1991.

Benjamin, Walter. "Epic Theater" and "The Work of Art in the Age of Mechanical Reproduction." *Illuminations.* Ed. Hannah Arendt. Trans. Harry Zohn. New York: Schocken, 1969. 149–56 and 219–54.

Bernaerts, Lars, Dirk De Geest, Luc Herman, and Bart Vervaek. *Stories and Minds: Cognitive Approaches to Literary Narrative.* Lincoln: U of Nebraska P, 2013.

Bernstein, Mark. "Patterns of Hypertext." *Proceedings of Hypertext 98: The Ninth ACM Conference on Hypertext and Hypermedia.* Ed. Kaj Grønbæk, Elli Mylonas, and Frank Shipman III. N.p.: ACM, 1998. 21–29.

Biocca, Frank, and Ben Delaney. "Immersive Virtual Reality Technology." Biocca and Levy, *Communication,* 57–124.

Biocca, Frank, Taeyong Kim, and Mark R. Levy. "The Vision of Virtual Reality." Biocca and Levy, *Communication,* 3–14.

Biocca, Frank, and Mark R. Levy, eds. *Communication in the Age of Virtual Reality.* Hillsdale, NJ: Lawrence Erlbaum, 1995.

Birkerts, Sven. *The Gutenberg Elegies: The Fate of Reading in an Electronic Age.* New York: Fawcett Columbine, 1994.

Boal, Augusto. "Theatre of the Oppressed." *The New Media Reader.* Ed. Noah Wardrip-Fruin and Nick Montfort. Cambridge, MA: MIT Press, 2003 [1974]. 339–52.

Bode, Christoph, and Rainer Dietrich. *Future Narratives: Theory, Poetics, and Media-Historical Moment.* Berlin: Walter De Gruyter, 2013.

Boellstorff, Tom. *Coming of Age in* Second Life. Princeton: Princeton UP, 2008.

Bolter, Jay David. "Literature in the Electronic Writing Space." Myron C. Tuman, *Literacy Online: The Promise (and Perils) or Reading and Writing with Computers.* Pittsburgh: U of Pittsburgh P, 1992. 19–42.

———. *Writing Space: The Computer, Hypertext, and the History of Writing.* Hillsdale, NJ: Lawrence Erlbaum, 1991.

Bolter, Jay David, and Richard Grusin. *Remediation: Understanding New Media.* Cambridge, MA: MIT Press, 1999.

Bouillot, Daniel. *Relation du texte à l'image et au son dans le cadre d'une fiction littéraire interactive.* Doctoral thesis, Université de Grenoble, France, 2012.

Brennan, Susan. "Conversation as Direct Manipulation: An Iconoclastic View." Laurel, *Art of Computer Interface Design,* 393–404.

Brewer, William F. "The Nature of Narrative Suspense and the Problem of Rereading." Vorderer, Wulff, and Friedrichsen, *Suspense,* 107–27.

Bricken, Meredith. "Virtual Worlds: No Interface to Design." Benedikt, *Cyberspace,* 363–82.

Caillois, Roger. *Men, Play, and Games.* Trans. Meyer Burasch. New York: Free Press, 1961.

Caïra, Olivier. *Jeux de rôles: Les forges de la fiction.* Paris: CNRS Editions, 2007.

Caïra, Olivier, and Susana Pajares Tosca. "Role-Playing Games." Ryan, Emerson, and Robertson, *The Johns Hopkins Guide to Digital Media,* 433–37.

Calabrese, Omar. *Neo-Baroque: A Sign of the Times.* Trans. Charles Lambert. Princeton, NJ: Princeton UP, 1992.

Calleja, Gordon. *In-Game: From Immersion to Incorporation.* Cambridge, MA: MIT Press, 2011.

Carroll, Noël. "The Paradox of Suspense." Vorderer, Wulff, and Friedrichsen, *Suspense,* 71–92.

Cavazza, Marc, Jean-Luc Lugrin, David Pizzi, and Fred Charles. "*Madame Bovary* on the Holodeck: Immersive Interactive Storytelling." *Proceedings of the ACM Multimedia 2007.* Augsburg, Germany: ACM Press, 2007. Available at https://ive.scm.tees.ac.uk/?pID=7.

Chandra, Vikram. *Geek Sublime: The Beauty of Code, the Code of Beauty.* Minneapolis: Graywolf, 2014.

Chatman, Seymour. *Coming to Terms: The Rhetoric of Narrative in Fiction and Film.* Ithaca: Cornell UP, 1990.

Ciccoricco, David. *Reading Network Fiction.* Tuscaloosa: U of Alabama P, 2007.

Coleridge, Samuel Taylor. *Biographia Literaria.* London: J. M. Dent, 1975.

Coover, Robert. "Hyperfiction: Novels for the Computer." *New York Times Book Review* 29 (Aug. 1993): 1, 8–12.

———. "Literary Hypertext: The Passing of the Golden Age." 1999. www.nickm.com/vox/golden_age.html.

Crawford, Chris. *Chris Crawford on Interactive Storytelling.* Berkeley: New Riders, 2004.

Culler, Jonathan. *Structuralist Poetics.* Ithaca: Cornell UP, 1975.

Currie, Gregory. "Imagination and Simulation: Aesthetics Meets Cognitive Science." Davies and Stone, *Mental Simulation,* 151–69.

———. *The Nature of Fiction.* Cambridge: Cambridge UP, 1990.

Davenport, Glorianna. "The Care and Feeding of Users." *IEEE Multimedia* 4, no. 1 (1997): 8–13.

Davies, Martin, and Tony Stone. *Mental Simulation.* Oxford: Blackwell, 1995.

de Jong, Irene. "Epic." Herman, Jahn, and Ryan, *The Routledge Encyclopedia of Narrative Theory,* 138–40.

Deleuze, Gilles, and Félix Guattari. *A Thousand Plateaus: Capitalism and Schizophrenia.* Trans. Brian Massumi. Minneapolis: U of Minnesota P, 1987.

Derrida, Jacques. "Structure, Sign, and Play in the Discourse of the Human Sciences." *The Languages of Criticism and the Sciences of Man.* Ed. Richard Macksey and Eugenio Donato. Baltimore: Johns Hopkins UP, 1970. 247–65.

Ditlea, Steve. "False Starts Aside, Virtual Reality Finds New Roles." *New York Times,* 23 Mar. 1998, C3.

Doležel, Lubomír. *Heterocosmica: Fiction and Possible Worlds.* Baltimore: Johns Hopkins UP, 1998.

Douglas, J. Yellowlees. "How Do I Stop This Thing?" Landow, *Hyper/Text/Theory,* 159–88.

———. "Hypertext, Argument, and Relativism." Snyder, *Page to Screen,* 144–62.

Eagleton, Terry. *Literary Theory: An Introduction.* Oxford: Basil Blackwell, 1983.

Eco, Umberto. *The Open Work.* Cambridge, MA: Harvard UP, 1989.

———. "Possible Worlds and Text Pragmatics: 'Un Drame bien parisien.'" *The Role of the Reader: Explorations in the Semiotics of Texts.* Bloomington: Indiana UP, 1979. 200–260.

———. *Travels in Hyperreality: Essays.* Trans. William Weaver. San Diego: Harcourt Brace Jovanovich, 1986.

Eliade, Mircea. *Myths, Rites, Symbols: A Mircea Eliade Reader.* Ed. Wendell C. Beane and William G. Doty. Vol. 1. New York: Harper & Row, 1975.

Emerson, Lori. *Reading Writing Interfaces: From the Digital to the Bookbound.* Minneapolis: U of Minnesota, 2014.

Ensslin, Astrid. *Literary Gaming.* Cambridge, MA: MIT Press, 2014.

Eskelinen, Markku. *Cybertext Poetics: The Critical Landscape of New Media Literary Theory.* London: Continuum, 2012.

Everding, Gerry. "Readers Build Vivid Mental Simulations of Narrative Situations, Brain Scans Suggest." *Record* (Washington University), 12 Feb. 2009. http://news.wustl.edu/news/Pages/13383.aspx.

Fischlin, Daniel, and Andrew Taylor. "Cybertheater, Postmodernism, and Virtual Reality: An Interview with Toni Dove and Michael Mackenzie." *Science-Fiction Studies* 21 (1994): 1–23.

Fleischman, Suzanne. *Tense and Narrativity: From Medieval Performance to Modern Fiction.* Austin: U of Texas P, 1990.

Foucault, Michel. "What Is an Author?" *The Foucault Reader.* Ed. Paul Rabinow. New York: Pantheon, 1984. 101–20.

Fournel, Paul, and Jean-Pierre Enard. "The Theater Tree: A Combinatory Play." Motte, *Oulipo,* 159–62.

Frazier, Ian. "The Novel's Main Character." *New Yorker,* 5 Sept. 1994, 59–60.

Friedman, Ted. "Making Sense of Software: Computer Games and Interactive Textuality." Jones, *Cybersociety,* 73–89.

Friess, Regina. "Play and Narrative as Patterns of Meaning Construction: Theoretical Foundation and Empirical Evaluation of the User's Experience in Interactive Film." Spierling and Szilas, *Interactive Storytelling,* 108–13.

Fuller, Matthew, ed. *Software Studies: A Lexicon.* Cambridge, MA: MIT Press, 2008.

Furst, Lilian R. *All Is True: The Claims and Strategies of Realist Fiction.* Durham, NC: Duke UP, 1995.

Gaggi, Silvio. *From Text to Hypertext: Decentering the Subject in Fiction, Film, the Visual Arts, and Electronic Media.* Philadelphia: U of Pennsylvania P, 1997.

Garner, Stanton B., Jr. *Bodied Space: Phenomenology and Performance in Contemporary Drama.* Ithaca: Cornell UP, 1994.

Gass, William. "Representation and the War for Reality." *Habitations of the Word.* New York: Simon & Schuster, 1985. 73–112.

Gerrig, Richard J. *Experiencing Narrative Worlds: On the Psychological Activities of Reading.* New Haven: Yale UP, 1993.

———. "The Resiliency of Suspense." Vorderer, Wulff, and Friedrichsen, *Suspense,* 93–105.

Glassner, Andrew. *Interactive Storytelling: Techniques for 21st Century Fiction.* Natick, MA: A. K. Peters, 2004.

Goodman, Nelson. "Reality Remade." *Philosophy Looks at the Arts.* Ed. Joseph Margolis. Philadelphia: Temple UP, 1987. 283–306.
———. *Ways of Worldmaking.* Indianapolis: Hackett, 1978.
Hansen, Mark B. N. *Bodies in Code: Interfaces with Digital Media.* London: Routledge, 2006.
Harrigan, Pat, and Noah Wardrip-Fruin, eds. *Second Person: Role-Playing and Story in Games and Playable Media.* Cambridge, MA: MIT Press, 2007.
Hayles, N. Katherine. "Artificial Life and Literary Culture." Ryan, *Cyberspace,* 205–23.
———. "The Condition of Virtuality." *The Digital Dialectic: New Essays on New Media.* Ed. Peter Lunenfeld. Cambridge, MA: MIT Press, 1999. 68–94.
———. *Electronic Literature: New Horizons for the Literary.* Notre Dame, IN: U of Notre Dame P, 2008.
———. "Embodied Virtuality, or How to Put Bodies Back into the Picture." Moser and McLeod, *Immersed,* 1–28.
———. *How We Became Posthuman: Virtual Bodies in Cybernetics, Literature and Informatics.* Chicago: U of Chicago P, 1999.
———. *How We Think: Digital Media and Contemporary Technogenesis.* Chicago: U of Chicago P, 2012.
———. "Hyper and Deep Attention: The Generational Divide in Cognitive Modes." *Profession* (2007): 187–99.
Heal, Jane. "How to Think about Thinking." Davies and Stone, *Mental Simulation,* 33–52.
Heim, Michael. "The Erotic Ontology of Cyberspace." Benedikt, *Cyberspace,* 59–80.
———. *The Metaphysics of Virtual Reality.* New York: Oxford UP, 1993.
———. *Virtual Realism.* New York: Oxford UP, 1998.
Herman, David. *Story Logic: Problems and Possibilities of Narrative.* Lincoln: U of Nebraska P, 2002.
Herman, David, Manfred Jahn, and Marie-Laure Ryan. *The Routledge Encyclopedia of Narrative Theory.* London: Routledge, 2005.
Hillis, Ken. *Digital Sensations.* Minneapolis: U of Minnesota P, 1999.
———. "Virtual Reality." Ryan, Emerson, and Robertson, *The Johns Hopkins Guide to Digital Media,* 510–14.
Hintikka, Jaakko. "Modality as Referential Multiplicity." *Ajatus* 20 (1957): 49–64.
Huizinga, Johan. *Homo Ludens: A Study of the Play Element in Culture.* Boston: Beacon, 1955.
Ingarden, Roman. *The Literary Work of Art: An Investigation on the Borderlines of Ontology, Logic, and the Theory of Literature.* Trans. George G. Grabowicz. Evanston: Northwestern UP, 1973.
Iser, Wolfgang. *The Fictive and the Imaginary: Charting Literary Anthropology.* Baltimore: Johns Hopkins UP, 1993.
———. "The Play of the Text." *Prospecting: From Reader Response to Literary Anthropology.* Baltimore: Johns Hopkins UP, 1989. 249–60.
———. "The Reading Process." *Reader-Response Criticism: From Formalism to Poststructuralism.* Ed. Jane P. Tomkins. Baltimore: Johns Hopkins UP, 1980. 50–69.
Jameson, Fredric. *Postmodernism, or The Cultural Logic of Late Capitalism.* London: Verso, 1991.
Jenkins, Henry. *Convergence Culture: Where Old and New Media Collide.* New York: New York UP, 2006.
———. "Game Design as Narrative Architecture." Wardrip-Fruin and Harrigan, *First Person,* 118–30.
Jones, Steven B., ed. *Cybersociety: Computer-Mediated Communication and Community.* Thousand Oaks, CA: Sage, 1995.
Joyce, Michael. *Of Two Minds: Hypertext, Pedagogy, and Poetics.* Ann Arbor: U of Michigan P, 1995.

Juul, Jesper. *Half-Real: Videogames Between Real Rules and Fictional Worlds.* Cambridge, MA: MIT Press, 2011.
Keen, Suzanne. "A Theory of Narrative Empathy." *Narrative* 14, no. 3 (2006): 207–36.
———. *Empathy and the Novel.* Oxford: Oxford UP, 2007.
Keep, Christopher J. "The Disturbing Liveliness of Machines: Rethinking the Body in Hypertext Theory and Fiction." Ryan, *Cyberspace,* 164–81.
Kelso, Margaret Thomas, Peter Weyhrauch, and Joseph Bates. "Dramatic Presence." *Presence: Teleoperators and Virtual Environments* 2, no. 1 (1993): 1–15.
Kripke, Saul. "Semantical Considerations on Modal Logic." *Acta Philosophica Fennica* 16 (1963): 83–94.
Kroker, Arthur, and Michael A. Weinstein. *Data Trash: The Theory of the Virtual Class.* New York: St. Martin's, 1994.
Krzywinska, Tanya. 2008. "World Creation and Lore: World of Warcraft as Rich Text." Hilde G. Corneliussen and Jill Walker Rettberg, eds., *Digital Culture, Play and Identity.* Cambridge, MA: MIT Press, 2008. 123–41.
Kuzmičová, Anežka. "The Words and Worlds of Narrative: The Trade-Off Between Verbal Presence and Direct Presence in the Activity of Reading." Bernaerts et al., *Cognitive Literary Studies,* 107–28.
Lamarque, Peter, and Stein Haugen Olsen. *Truth, Fiction, and Literature.* Oxford: Clarendon, 1994.
Landow, George P. *Hypertext 2.0: The Convergence of Contemporary Critical Theory and Technology.* Baltimore: Johns Hopkins UP, 1997.
———, ed. *Hyper/Text/Theory.* Baltimore: Johns Hopkins UP, 1994.
Langer, Susanne K. *Feeling and Form: A Theory of Art.* New York: Scribner's, 1953.
Lanham, Richard. *The Electronic Word: Democracy, Technology, and the Arts.* Chicago: U of Chicago P, 1993.
Lanier, Jaron, and Frank Biocca. "An Insider's View of the Future of Virtual Reality." *Journal of Communications* 42, no. 4 (1992): 150–72.
Lasko-Harvill, Ann. "Identity and Mask in Virtual Reality." *Discourse* 14, no. 2 (1992): 222–34.
Laurel, Brenda. "Art and Activism in VR." *Wide Angle* 15, no. 4 (1993): 13–21.
———. *Computers as Theatre.* Menlo Park, CA: Addison Wesley, 1991.
———. *Utopian Entrepreneur.* Cambridge, MA: MIT Press, 2001.
———, ed. *The Art of Computer Interface Design.* Reading, MA: Addison-Wesley, 1990.
Laurel, Brenda, Rachel Strickland, and Rob Tow. "Placeholder: Landscape and Narrative in Virtual Environments." *Computer Graphics* 28, no. 2 (1994): 118–26.
Lazzaro, Nicole. "Why We Play Games: Four Keys to More Emotions Without Story." www.xeodesign.com/xeodesign_whyweplaygames.pdf.
Lévy, Pierre. *Becoming Virtual: Reality in the Digital Age.* Trans. Robert Bonono. New York: Plenum Trade, 1998.
———. *L'Idéographie dynamique.* Paris: La Découverte, 1991.
Lewis, David. *On the Plurality of Worlds.* Cambridge: Blackwell, 1986.
———. "Truth in Fiction." *American Philosophical Quarterly* 15 (1978): 37–46.
Liestøl, Gunnar. "Wittgenstein, Genette, and the Reader's Narrative in Hypertext." Landow, *Hyper/Text/Theory,* 87–120.
Liu, Alan. *The Laws of Cool: Knowledge Work and the Culture of Information.* Chicago: U of Chicago P, 2004.
Loeffler, Carl Eugene, and Tim Anderson, eds. *The Virtual Reality Casebook.* New York: Van Nostrand Reinhold, 1994.
Lotman, Juri. *The Structure of the Artistic Text.* Trans. G. Lehnhoff and R. Vroom. Ann Arbor: U of Michigan P, 1970.

Lugrin, Jean-Luc, Fred Charles, Marc Cavazza, Marc Le Renard, Jonathan Freeman, and Jane Lessiter. "Are Immersive FPS Games Enjoyable." *Proceedings of the 18th ACM Symposium on Virtual Reality Software and Technology* (VRST '12). New York, NY: ACM, 2012. 199–200.

Lyotard, Jean-François. *The Postmodern Condition: A Report on Knowledge.* Trans. Geoff Bennington and Brian Massumi. Minneapolis: U of Minnesota P, 1991.

Maître, Doreen. *Literature and Possible Worlds.* London: Middlesex Polytechnic Press, 1983.

Makaryk, Irena K., ed. *Encyclopedia of Contemporary Literary Theory: Approaches, Scholars, Terms.* Toronto: U of Toronto P, 1993.

Maloney, Judith. "Fly Me to the Moon: A Survey of American Historical and Contemporary Simulation Entertainments." *Presence: Teleoperators and Virtual Environments* 6, no. 5 (1997): 565–80.

Mandler, Jean, and Nancy Johnson. "Remembrance of Things Parsed: Story Structure and Recall." *Cognitive Psychology* 9 (1977): 111–51.

Manovich, Lev. *The Language of New Media.* Cambridge, MA: MIT Press, 2001.

Marr, Lisa. Appendix to David Ciccoricco, *Refiguring Minds in Narrative Media.* Lincoln: U of Nebraska P, 2015.

Martin, Wallace. *Recent Theories of Narrative.* Ithaca: Cornell UP, 1986.

Martínez-Bonati, Félix. *Fictive Discourse and the Structures of Literature: A Phenomenological Approach.* Trans. Philip W. Silver. Ithaca: Cornell UP, 1981.

Mateas, Michael. "A Preliminary Poetics for Interactive Drama and Games." Wardrip-Fruin and Harrigan, *First Person,* 19–33.

———. "Weird Languages." Fuller, *Software Studies: A Lexicon,* 267–75.

McGonigal, Jane. *Reality Is Broken.* London: Penguin, 2011.

McHale, Brian. *Postmodernist Fiction.* London: Methuen, 1987.

McLuhan, Marshall. *Essential McLuhan.* Ed. Eric McLuhan and Frank Zingrone. New York: Basic Books, 1996.

Mechner, Jordan. "*The Sands of Time*: Crafting a Video Game Story." Harrigan and Wardrip-Fruin, *Second Person,* 111–20.

Merleau-Ponty, Maurice. *The Phenomenology of Perception.* Trans. Colin Smith. London: Routledge & Kegan Paul, 1962.

———. *The Primacy of Perception.* Ed. James M. Edie. Chicago: Northwestern UP, 1964.

Milam, David, Magy Seif El-Nasr, and Ron Wakkary. "Looking at the Interactive Narrative Experience through the Eye of the Participants." Spierling and Szilas, *Interactive Storytelling,* 96–107.

Miller, J. Hillis. *Topographies.* Stanford, CA: Stanford UP, 1995.

Montfort, Nick. "Interactive Fiction as 'Story,' 'Game,' 'Storygame,' 'Novel,' 'World,' 'Literature,' 'Puzzle,' 'Problem,' 'Riddle,' and 'Machine.'" Wardrip-Fruin and Harrigan, *First Person,* 310–17.

———. "Obfuscated Code." Fuller, *Software Studies: A Lexicon.* 193–99.

Moretti, Franco. *Distant Reading.* London: Verso, 2013.

Moser, Mary Anne, and Douglas McLeod, eds. *Immersed in Technology: Art and Virtual Environments.* Cambridge, MA: MIT Press, 1996.

Mosher, Harold F. "Toward a Poetics of 'Descriptized' Narration." *Poetics Today* 12, no. 3 (1991): 425–46.

Motte, Warren F. *Playtexts: Ludics in Contemporary Literature.* Lincoln: U of Nebraska P, 1995.

———, ed. and trans. *Oulipo: A Primer of Potential Literature.* Lincoln: U of Nebraska P, 1986.

Moulthrop, Stuart. "Rhizome and Resistance: Hypertext and the Dreams of a New Culture." Landow, *Hyper/Text/Theory,* 299–322.

Murray, Janet H. *Hamlet on the Holodeck: The Future of Narrative in Cyberspace.* New York: Free Press, 1997.

———. *Inventing the Medium: Principles of Interaction Design as a Cultural Practice.* Cambridge, MA: MIT Press, 2012.
Ndalianis, Angela. *Neo-Baroque Aesthetics and Contemporary Entertainment.* Cambridge, MA: MIT Press, 2004.
Nell, Victor. *Lost in a Book: The Psychology of Reading for Pleasure.* New Haven: Yale UP, 1988.
Nietzsche, Friedrich. *The Birth of Tragedy and the Case of Wagner.* Trans. Walter Kaufmann. New York: Random House, 1967.
Nitsche, Michael. *Video Game Space.* Cambridge, MA: MIT Press, 2008.
Novak, Marcos. "Liquid Architecture in Cyberspace." Benedikt, *Cyberspace,* 225–54.
Nunes, Mark. "Virtual Topographies: Smooth and Striated Cyberspace." Ryan, *Cyberspace,* 61–77.
Oatley, Keith. *Such Stuff as Dreams: The Psychology of Fiction.* Malden, MA: Wiley-Blackwell, 2011.
———. "Why Fiction May be Twice as True as Fact: Fiction as Cognitive and Emotional Simulation." *Review of General Psychology* 3, no. 2 (1989): 101–17.
Ong, Walter J. *Orality and Literacy: The Technologizing of the Word.* London: Methuen, 1982.
Oren, Tim. "Designing a New Medium." Laurel, *Art of Computer Interface Design,* 467–79.
Paul, Christiane. *Digital Art.* London: Thames and Hudson, 2003.
Pavel, Thomas. *Fictional Worlds.* Cambridge, MA: Harvard UP, 1986.
Pearce, Celia. *The Interactive Book.* Indianapolis: Macmillan Technical, 1997.
Penny, Simon. "Virtual Reality as the Completion of the Enlightenment." Loeffler and Anderson, *Virtual Reality Casebook,* 199–213.
Phelan, James. *Reading People, Reading Plots: Character, Progression, and the Interpretation of Narrative.* Chicago: U of Chicago P, 1989.
Pimentel, Ken, and Kevin Teixeira. *Virtual Reality: Through the New Looking Glass.* New York: Intel/Windcrest McGraw Hill, 1993.
Poole, Steven. *Trigger Happy: The Inner Life of Video Games.* New York: Arcade, 2000.
Popper, Frank. *Art—Action and Participation.* New York: New York UP, 1975.
Poster, Mark. "Theorizing Virtual Reality: Baudrillard and Derrida." Ryan, *Cyberspace,* 42–60.
Pratt, Mary Louise. *Toward a Speech Act Theory of Literary Discourse.* Bloomington: Indiana UP, 1977.
Propp, Vladimir. *Morphology of the Folk-tale.* Trans. Laurence Scott. Rev. Louis A. Wagner. Austin: U of Texas P, 1968 [1928].
Rees, Gareth. "Tree Fiction on the World Wide Web." http://lucilia.ebc.ee/~enok/tree-fiction.html.
Reid, Elizabeth. "Virtual Worlds: Culture and Imagination." Jones, *Cybersociety,* 164–83.
Rheingold, Howard. *Virtual Reality.* New York: Simon & Schuster, 1991.
Ronen, Ruth. *Possible Worlds in Literary Theory.* Cambridge: Cambridge UP, 1994.
Roose-Evans, James. *Experimental Theatre: From Stanislavsky to Peter Brooks.* London: Routledge, 1989.
Rotman, Brian. "Thinking Dia-Grams: Mathematics, Writing, and Virtual Reality." *South Atlantic Quarterly* 94, no. 2 (1995): 389–415.
Rousset, Jean. *L'Intérieur et l'extérieur: Essais sur la poésie et sur le théâtre au XVIIe siècle.* Paris: Corti, 1968.
Ruston, Scott. "Location-Based Narrative." Ryan, Emerson, and Robertson, *The Johns Hopkins Guide to Digital Media,* 318–21.
Ryan, Marie-Laure. *Avatars of Story.* Minneapolis: U of Minnesota P, 2006.
———, ed. *Cyberspace Textuality: Computer Technology and Literary Theory.* Bloomington: Indiana UP, 1999.
———. *Possible Worlds, Artificial Intelligence, and Narrative Theory.* Bloomington: Indiana UP, 1991.
———. "Possible Worlds in Recent Literary Theory." *Style* 26, no. 4 (1992): 528–53.

———. "Toward a Definition of Narrative." *The Cambridge Companion to Narrative.* Ed. David Herman. Cambridge: Cambridge UP, 2007. 22–35.

Ryan, Marie-Laure, Lori Emerson, and Benjamin Robertson, eds. *The Johns Hopkins Guide to Digital Media.* Baltimore: Johns Hopkins UP, 2014.

Seaman, Bill. "Interactive Text and Recombinant Poetics—Media-Element Field Explorations." Wardrip-Fruin and Harrigan, *First Person,* 227–34.

Searle, John. "The Logical Status of Fictional Discourse." *New Literary History* 6 (1974/75): 319–32.

———. *Speech Acts: An Essay in the Philosophy of Language.* Cambridge: Cambridge UP, 1969.

Semino, Elena. *Language and World Creation in Poems and Other Texts.* London: Longman, 1997.

Sheridan, Thomas B. "Musings on Telepresence and Virtual Presence." *Presence: Teleoperators and Virtual Environments* 1, no. 1 (1992): 120–25.

Simanowski, Roberto. *Digital Art and Meaning: Reading Kinetic Poetry, Text Machines, Mapping Art, and Interactive Installations.* Minneapolis: U of Minnesota P, 2011.

———. "Double Coding." Paper presented at the conference "Remediating Literature," Utrecht, Netherlands, July 2007.

Smith, Barbara Herrnstein. *On the Margins of Discourse.* Chicago: U of Chicago P, 1978.

Snyder, Ilana. "Beyond the Hype: Reassessing Hypertext." Snyder, *Page to Screen,* 125–43.

———. *Hypertext: The Electronic Labyrinth.* Melbourne: Melbourne UP, 1996.

———, ed. *Page to Screen: Taking Literacy into the Electronic Era.* London: Routledge, 1998.

Spariosu, Mihai I. *Dionysus Reborn: Play and the Aesthetic Dimension in Modern Philosophical and Scientific Discourse.* Ithaca: Cornell UP, 1989.

Speer, Nicole K., Jeremy R. Reynolds, Kheena M. Swallow, and Jeffrey M. Zacks. "Reading Stories Activates Neural Representations of Perceptual and Motor Experiences." *Psychological Science* 20 (2009): 989–99.

Spierling, Ulrike, and Nicolas Szilas, eds. *Interactive Storytelling: First Joint Conference on Interactive Digital Storytelling, Erfurt, Germany 2008 Proceedings.* Berlin: Springer, 2008.

Starr, G. Gabrielle. "Multisensory Imagery." Zunshine, *Introduction to Cognitive Cultural Studies,* 275–91.

Stefans, Brian Kim. "Against Desire: Excess, Disgust and the Sign in Electronic Literature." *Electronic Book Review.* http://electronicbookreview.com/thread/electropoetics/disgust.

Stenger, Nicole. "Mind Is a Leaky Rainbow." Benedikt, *Cyberspace,* 49–58.

Sternberg, Meir. "Proteus in Quotation-Land: Mimesis and the Forms of Reported Discourse." *Poetics Today* 3, no. 2 (1982): 107–56.

———. Telling in Time (II): Chronology, Teleology, Narrativity." *Poetics Today* 13, no. 3 (1992): 463–541.

Steuer, Jonathan. "Defining Virtual Reality: Dimensions Determining Telepresence." *Journal of Communications* 42, no. 4 (1992): 73–93.

Stich, Stephen, and Shaun Nichols. "Second Thoughts on Simulation." Davies and Stone, *Mental Simulation,* 87–108.

Stone, Allucquère Rosanne. "Will the Real Body Please Stand Up: Boundary Stories about Virtual Culture." Benedikt, *Cyberspace,* 81–118.

Suits, Bernard. "The Detective Story: A Case Study of Games in Literature." *Canadian Review of Comparative Literature* 12, no. 2 (1985): 200–219.

———. *The Grasshopper: Games, Life, and Utopia.* Toronto: U of Toronto P, 1978.

Szilas, Nicolas. "Interactive Drama on Computer: Beyond Linear Narrative." *Narrative Intelligence: Papers from the 1999 AAAI Fall Symposium.* Technical Report FS-99-01. Menlo Park: AAAI (American Association for Artificial Intelligence) Press, 1999.

———. "A New Approach to Interactive Drama: From Intelligent Characters to an Intelligent Virtual Narrator." http://www.aaai.org/Papers/Symposia/Spring/2001/SS-01-02/SS01-02-016.pdf.

Theall, Donald. *Beyond the Word: Reconstructing Sense in the Joyce Era of Technology, Culture, and Communication.* Toronto: U of Toronto P, 1995.
Therrien, Carl. "Interface." Ryan, Emerson, and Robertson, *The Johns Hopkins Guide to Digital Media,* 304–8.
Tooley, Michael. *Time, Tense, and Causation.* Oxford: Clarendon, 1997.
Truffaut, François. *Hitchcock.* New York: Simon & Schuster, 1967.
Tuan, Yi-Fu. *Space and Place: The Perspective of Experience.* Minneapolis: U of Minnesota P, 1977.
Turkle, Sherry. *Life on the Screen: Identity in the Age of the Internet.* New York: Simon & Schuster, 1995.
Ulmer, Gregory. "A Response to Twelve Blue by Michael Joyce." *Postmodern Culture* 8, no. 1 (1998). http://muse.jhu.edu/journals/postmodern culture/v008/8.1ulmer.html.
Van Looy, Jan, and Jan Baetens, eds. *Close Reading New Media: Analyzing Electronic Literature.* Leuven: Leuven UP, 2003.
Vermeule, Blakey. *Why Do We Care About Literary Characters?* Baltimore: Johns Hopkins UP, 2010.
Vorderer, Peter, Hans J. Wulff, and Mike Friedrichsen, eds. *Suspense: Conceptualizations, Theoretical Analyses, and Empirical Explorations.* Mahwah, NJ: Lawrence Erlbaum, 1996.
Vosmeer, Mirjam, and Ben Schouten. "Interactive Cinema: Engagement and Interaction." *2014 ICIDS Proceedings,* ed. A. Mitchell et al. Springer International Proceedings, 2014. 140–47.
Walser, Randall. "Spacemakers and the Art of the Cyberspace Playhouse." *Mondo 2000,* no. 2 (Summer 1990): 60–61.
Walsh, Richard. "Why We Wept for Little Nell: Character and Emotional Involvement." *Narrative* 5, no. 3 (1997): 306–21.
Walton, Kendall. *Mimesis as Make-Believe: On the Foundations of the Representational Arts.* Cambridge, MA: Harvard UP, 1990.
———. "Spelunking, Simulation, and Slime: On Being Moved by Fiction." *Emotion and the Arts.* Ed. Mette Hjort and Sue Laver. Oxford: Oxford UP, 1997. 37–49.
Wardrip-Fruin, Noah, and Pat Harrigan, eds. *First Person: New Media as Story, Performance, and Game.* Cambridge, MA: MIT Press, 2004.
Welty, Eudora. "Place in Fiction." *South Atlantic Quarterly* 55 (1956): 57–72.
Wertheim, Margaret. *The Pearly Gates of Cyberspace: A History of Space from Dante to the Internet.* New York: Norton, 1999.
Wilson, R. Rawdon. *In Palamedes' Shadow: Explorations in Play, Game, and Narrative Theory.* Boston: Northeastern UP, 1990.
Wittgenstein, Ludwig. *Philosophical Investigations.* Trans. G. E. M. Anscombe. New York: Macmillan, 1968.
Zahorik, Pavel, and Rick L. Jenison. "Presence as Being-in-the World." *Presence* 7, no 1 (1998): 78–89.
Zhai, Philip. *Get Real: A Philosophical Adventure in Virtual Reality.* New York: Rowman & Littlefield, 1999.
Žižek, Slavoj. *Tarrying with the Negative: Kant, Hegel, and the Critique of Ideology.* Durham, NC: Duke UP, 1993.
Zubin, David A., and Lynne E. Hewitt. "The Deictic Center: A Theory of Deixis in Narrative." *Deixis in Narrative: A Cognitive Science Perspective.* Ed. Judith F. Duchan, Gail A. Bruder, and Lynne E. Hewitt. Hillsdale, NJ: Lawrence Erlbaum, 1995. 129–55.
Zunshine, Lisa. *Why We Read Fiction: Theory of Mind and the Novel.* Columbus: Ohio State UP, 2006.
———, ed. *Introduction to Cognitive Cultural Studies.* Baltimore: Johns Hopkins UP, 2010.
Zwaan, Rolf A. "Situation Model." Herman, Jahn, and Ryan, *The Routledge Encyclopedia of Narrative Theory,* 534–35.

Index